Principles of Astrophotonics

Advanced Textbooks in Physics

ISSN: 2059-7711

The *Advanced Textbooks in Physics* series explores key topics in physics for MSc or PhD students.

Written by senior academics and lecturers recognised for their teaching skills, they offer concise, theoretical overviews of modern concepts in the physical sciences. Each textbook comprises of 200–300 pages, meaning content is specialised, focussed and relevant.

Their lively style, focused scope and pedagogical material make them ideal learning tools at a very affordable price.

Published

Principles of Astrophotonics
by Simon Ellis, Joss Bland-Hawthorn and Sergio Leon-Saval

Topics in Statistical Mechanics (Second Edition)
by Brian Cowan

Physics of Electrons in Solids: Design and Applications
by Jean-Claude Tolédano

Astronomical Spectroscopy: An Introduction to the Atomic and Molecular Physics of Astronomical Spectroscopy (Third Edition)
by Jonathan Tennyson

A Guide to Mathematical Methods for Physicists: Advanced Topics and Applications
by Michela Petrini, Gianfranco Pradisi & Alberto Zaffaroni

Quantum States and Scattering in Semiconductor Nanostructures
by Camille Ndebeka-Bandou, Francesca Carosella & Gérald Bastard

An Introduction to Particle Dark Matter
by Stefano Profumo

Studying Distant Galaxies: A Handbook of Methods and Analyses
by François Hammer, Mathieu Puech, Hector Flores & Myriam Rodrigues

Trapped Charged Particles: A Graduate Textbook with Problems and Solutions
edited by Martina Knoop, Niels Madsen & Richard C Thompson

Advanced Textbooks in Physics

PRINCIPLES OF ASTROPHOTONICS

Simon Ellis
Macquarie University, Australia

Joss Bland-Hawthorn
The University of Sydney, Australia

Sergio Leon-Saval
The University of Sydney, Australia

NEW JERSEY • LONDON • SINGAPORE • BEIJING • SHANGHAI • HONG KONG • TAIPEI • CHENNAI • TOKYO

Published by

World Scientific Publishing Europe Ltd.

57 Shelton Street, Covent Garden, London WC2H 9HE

Head office: 5 Toh Tuck Link, Singapore 596224

USA office: 27 Warren Street, Suite 401-402, Hackensack, NJ 07601

Library of Congress Cataloging-in-Publication Data
Names: Ellis, Simon C., author. | Bland-Hawthorn, J. (Joss), author. | Leon-Saval, Sergio, author.
Title: Principles of astrophotonics / Simon Ellis, Macquarie University, Australia,
Joss Bland-Hawthorn, The University of Sydney, Australia,
Sergio Leon-Saval, The University of Sydney, Australia.
Description: New Jersey : World Scientific Publishing, [2023] | Series: Advanced textbooks in physics, 2059-7711 | Includes bibliographical references and index.
Identifiers: LCCN 2022030933 | ISBN 9781800613256 (hardcover) |
ISBN 9781800613355 (paperback) | ISBN 9781800613263 (ebook for institutions) |
ISBN 9781800613270 (ebook for individuals)
Subjects: LCSH: Astronomical instruments. | Photonics.
Classification: LCC QB86 .E45 2023 | DDC 621.36/5015222--dc23/eng20221013
LC record available at https://lccn.loc.gov/2022030933

British Library Cataloguing-in-Publication Data
A catalogue record for this book is available from the British Library.

For any available supplementary material, please visit
https://www.worldscientific.com/worldscibooks/10.1142/Q0391#t=suppl

Desk Editors: Nimal Koliyat/Adam Binnie/Shi Ying Koe

Typeset by Stallion Press
Email: enquiries@stallionpress.com

In loving memory of Maz

PREFACE

Astrophotonics is the application of photonics to astronomical instrumentation. This rapidly developing field has reached a stage where many prototype devices are now being tested on sky, and the first fully-fledged instruments incorporating photonic devices are now being used for observations. The field is thus transitioning from one of instrumental research and development to mainstream observational astrophysics. This book is intended to communicate the current status, potential, and future possibilities of the field to the wider astronomical, optics and photonics communities during this period of transition.

The rise of astrophotonics marks a significant departure from the traditional approach to astronomical instrumentation. Astronomical instruments typically employ bulk optics, such as lenses, mirrors, diffraction gratings and filters to manipulate beams of light that propagate in free-space between the optics. Astrophotonic instruments replace such bulk optics with devices embedded in optical fibres and waveguides, such that the manipulation of light takes place within the waveguides themselves. This allows improvements in efficiency, functionality and scalability of astronomical instruments.

Thus, astrophotonics is a whole new approach to instrumentation, which seeks to employ new ways to manipulate light within waveguides to provide new and better instruments. At the heart of astrophotonics therefore, and fundamental to it, is our understanding and treatment of light itself.

Our perception of the nature of light has evolved over the centuries. In the early 19th century, Thomas Young and Augustin Fresnel established that light is a wave phenomenon, and that these waves are able to overlap, much like ripples on the surface of a pond. The principle of superposition was an important step because it meant that wave displacements sum algebraically, such that waves are able to combine constructively or destructively. Already by 1820, the phenomenon of the interference of light was well understood.

But a deeper understanding had to wait another 40 years until what Dirac, Feynman and many others have argued as one of the most remarkable milestones in the evolution of physics, on a par with those we associate with Newton and Einstein. This was Maxwell's theory of electrodynamics where light waves are manifestations of oscillating electric and magnetic fields. Glauber (2006) has argued that there have been no fundamental additions to Maxwell's equations over the past 150 years. These equations still serve as the basis for the design and analysis of all optical instruments and telescopes to date.

The laws governing the granularity of light follow a parallel track starting with Planck at the close of the 19th century. His analysis of the blackbody spectrum led to an empirical formula that he set out in his famous 1900 paper. The formula was accepted immediately, which led the scientific community to ask what was the physical basis for such a formula. He needed a model of a blackbody that could efficiently interact with an external reservoir, a notion that already existed in thermodynamics. Along much the same lines that statistical or quantum mechanics are taught today (Schroeder 2000), he devised a mechanical system made up of one-dimensional oscillators that could absorb and emit radiation. Maxwell's equations tell us how a charge interacts with the electromagnetic field. This forced Planck to a remarkable insight: the formula followed if the distribution of permitted energies for an oscillator with frequency ν was discretised like the rungs of a ladder, i.e. oscillators with energies $nh\nu$ given $n = 0, 1, 2, \ldots$. Planck referred to the fundamental unit as the quantum of energy $h\nu$, a concept that troubled him greatly because, after all, waves observed in nature appear to exist at any frequency and therefore form a continuous distribution.

Five years on, Einstein recalled Hertz's 1887 experiment of the photoelectric effect. He also noticed that the entropy of Planck's radiative distribution resembled the entropy of a perfect gas of free particles with energy $h\nu$. Thus, over several papers in 1905, Einstein's annus mirabilis, he introduced the fundamental basis for the existence of photons. Interestingly, the quantised nature of light was treated with some scepticism until Compton's discovery in 1922 that X-ray quanta were scattered by electrons. Indeed, by 1926, Lewis had coined the term 'photons' for these mysterious particles of light.

In this book, it is sometimes necessary to treat light classically (pre-1900) or in the modern context of the quantum field. Crucially, Dirac showed that light and charges exist within a vacuum (space) that must be treated as an active dynamical system. The energy state of the vacuum cannot be zero because this is ruled out by Heisenberg's uncertainty principle. Thus, the energy levels of Planck's ladder are not $nh\nu$ ($n = 0, 1, 2, \ldots$), but $\left(n + \frac{1}{2}\right) h\nu$. The half-quantum of energy ensures that oscillations can never be permanently at rest, therefore the vacuum is 'alive' with a roiling ocean of weak electromagnetic fields. Dirac's vacuum is the basis of many physical phenomena, including atomic spontaneous emission, the Casimir effect, the Lamb shift and Hawking radiation. Indeed, the vacuum fluctuations excite background noise in quantum amplifiers that have important applications in astrophotonics.

For an introductory text, we avoid the use the language of quantum field theory, e.g. the high-order correlations of photons ($g^{(1)}, g^{(2)}$, etc.). All of classical optics and photonics can be cast in this 'light', for example, the Young's slit experiment and grating spectroscopy. We recognise that quantum optics is experiencing a resurrection (e.g. Guerin *et al.* 2018) since its emergence in the early 1950s in the Hanbury Brown–Twiss (HBT) experiments. With the advent of extremely large telescopes, there are important respects in which quantum optics and quantum photonics are being realised within astrophotonics (Bland–Hawthorn *et al.* 2021), but we consider these to be advanced topics for another book. We touch on these developments in the final chapter.

For any text, it is imperative that there be a consistent language in describing classical optics and photonics. In view of the historical introduction, it is sometimes convenient to treat light as ray bundles travelling in straight lines. This was how optical design was conducted since the invention of the telescope by Hans Lippershey until the advent of wave equations. This approach is still used in the early stages of optical design, not least because it is more easily rendered. But final designs are typically from treating light as a propagating wave because this is demanded by the optical transfer function (i.e. complete mathematical description) of the system. Thus, widely used optics packages like `Zemax` make the distinction between geometric optics and physical wave optics, and both options can be used interchangeably for the same optical design.

The distinction between the interference of waves and photons troubles the majority of scientists. Does it matter where on the wavefront the photon is being carried towards the telescope? To see interference fringes from a nearby star, does that require identical photons from the star to arrive at both telescopes simultaneously, and so on. Such concerns troubled Hanbury Brown and Twiss in the early years of their work (Hanbury Brown and Twiss 1956). However, it is not the photons but the probability amplitudes that interfere within a correlation time given by the spectral bandpass of the observation (Purcell 1956). This is the basis for Feynman's statement that we only need to understand Young's elegant slit experiment to appreciate the (quantum) nature of light.

By the late 1950s, another important event occurred in the story of light — the laser was born (Javan *et al.* 1961). While Planck's blackbody can be considered as an incoherent source arising from a current of free charges, the laser was something altogether different. This was a polarisation current of bound charges oscillating perpendicular to the tube's axis, generating a (Poisson) distribution of coherent photons that were statistically independent. The laser is the bedrock of photonics and of course it powers the telecommunications industry. The laser's properties and its applications are explored throughout this book.

Even though we do not consider quantum optics and quantum photonics in any depth, there are important instances where we treat light discretely rather than as a continuous wave. This is necessary, for example, to clarify the distinction between wave (phase) interferometry and intensity interferometry. Moreover, astronomy is the study of faint sources where we are frequently in the photon-counting limit.

To make the connection with celestial sources, to a fair approximation, most stars can be considered as blackbodies. The emitted radiation is highly incoherent given the nature of the hot atmospheric layers. As seen from Earth, an unresolved star is a coherent source. In the star's atmosphere, the oscillating charges are dipoles that emit radiation within a solid angle. The light waves are stretched and flattened as they propagate over vast distances. Much like throwing a handful of pebbles into a pond, at the point of impact, the ripples are an incoherent jumble, but after a time a coherent pattern emerges. Thus, we can consider stars observed in a monochromatic bandpass to be a crude, uncollimated laser source by virtue of the van Cittert–Zernike theorem.

But our analogy breaks down when a star can be 'resolved' by the observing telescope or array of telescopes, i.e. when it subtends a finite, measurable angle as seen by the observer. This is what is revealed by the Hanbury Brown–Twiss experiment. The second-order correlation coefficient $g^{(2)}$ reveals that the photon events detected at two suitably spaced telescopes are bunched ($g^{(2)} = 2$ for an ideal system). This requires a *mixture* of pure coherent states (i.e. incoherent radiation in a restricted spectral band with finite width, not monochromatic) for the correlation to be seen. If, by some remarkable feat, we were observing a laser source at the star's distance, there would be no such correlation ($g^{(2)} = 1$). If the laser were replaced by a fluorescent source, the photons would be anti-bunched such that $g^{(2)} < 1$. Thus, the granular nature of light provides unique information about the source that cannot be determined from a continuous wave. With the advent of the extremely large telescopes, we would argue that quantum optics and quantum photonics are the next 'wave' in astrophotonics, a topic we explore in the final chapter of the book.

These days, a clear distinction is made between optics (e.g. lenses, mirrors) and photonics (optical fibres, planar waveguides). Photonics has been described as moulding the flow of light (Bland–Hawthorn and Kern 2012). An aspect of our discussion of photonics, which applies to both continuous waves and discrete photons, is the concept of modes. This language will be familiar to students of acoustics (e.g. the modes of vibration of a violin string) and to students of quantum mechanics (e.g. mode occupation for a given distribution). Modes are the solutions (or eigenfunctions) of a wave equation that describes how waves propagate within a given system. These are orthogonal such that they do not interfere with each other. Photons propagating within the same mode cannot be distinguished by their energy $h\nu$. The number of photons n per mode is tiny for a thermal source ($n \ll 1$) but very large for a laser source ($n \gg 1$). More fundamentally, while all photons are coherent, only those described by the same mode (and same polarisation) can interfere.

For our purposes, there are two types of modes: *spatial modes* transverse to the direction of propagation (beam cross section, beam divergence), and *temporal modes* aligned with the direction of propagation (time, frequency). In experimental photonics, there are many kinds of beams that are shaped for a range of purposes described by different modal families. Modes or eigenfunctions are not abstract concepts, any more than standing waves within a musical instrument. Reflected laser light from the body of a cello reveals a rich tapestry of standing waves (e.g. Fleischer 2011). No less impressive, Midwinter (1975) has photographed the many modes that propagate within a multimode fibre. How this is even possible is an exercise for the reader.

A proper treatment of spatial modes preserves the system's étendue, i.e. conservation of the beam cross section and solid angle (divergence) product. The exact shapes of the spatial modes depend on the boundary condition, whether circular, planar or otherwise. Modal analysis simplifies the analysis of more complex structures like metamaterials and photonic crystal fibres. A mode's shape can be altered by a passive optical element, but the number of photons per mode can never be increased (Daendliker 2000). Moreover, the 'phase space volume' defined by the transverse shape (x, y) and momentum (z) of the photon is preserved also.

Of equal importance, a full treatment of temporal modes reveals the central role of the correlation length (or time) of the system. The central role of the Wiener–Kitchin theorem describes the deep relationship between the temporal coherence function and the source's spectrum (power spectral density). This reveals that the coherence time of any source is inversely related to the observed spectral bandpass.

The ideas presented here on the treatment and understanding of the nature of light are at the core of astrophotonics; new ways of manipulating light to improve our measurements and understanding of the Universe naturally require a full and thorough understanding of light itself. Astrophotonics often requires a full electromagnetic and modal treatment, or even a quantum treatment of light, whereas for classical astronomical optics geometric approximations often suffice. These ideas are revisited and expanded upon throughout the book.

ABOUT THE AUTHORS

Simon Ellis is an Associate Professor at Macquarie University where he is the head of the Instrument Science Group, which is responsible for research and development in astronomical instrumentation and technologies. He has over 20 years' experience of research in astronomy and astrophysics, including instrumentation research and development, the evolution of galaxies and clusters of galaxies, and the astrophysics of positronium. He has played a leading role in the development of astrophotonics, most notably in the field of OH suppression. He was the commissioning scientist for the world's first two instruments to use fibre Bragg gratings for OH suppression. He has published foundational papers in the use of silicon photonics for astronomy. He has been invited to speak on astrophotonics at major photonics and astronomy conferences, and was guest editor for a feature issue on astrophotonics in the *Journal of the Optical Society of America B*.

Joss Bland-Hawthorn is a Laureate Professor of Physics, and Director of the Sydney Institute for Astronomy in The University of Sydney. He has more than 30 years of experience in astrophysics and astronomical instrumentation. In 2000, he founded the field of astrophotonics, now a major subfield of photonics. With Birks and Leon-Saval, he developed (and named) the photonic lantern, a device that is finding wider use in science and industry. He has developed and introduced many other enabling technologies, including OH suppressing fibre Bragg gratings (with Ellis), hexabundles, photonic microspectrograph concepts, tunable filters, and so forth.

Sergio Leon-Saval is an Associate Professor at the School of Physics in The University of Sydney where he is now Director of the Sydney Astrophotonics Instrumentation Laboratory (SAIL), and Deputy Director of the Institute of Photonics and Optical Science (IPOS). He has more than 16 years of experience in the research area of photonics, and made breakthrough contributions in the fields of speciality optical fibres and astrophotonics, including the co-invention and pioneering the development of the photonic lantern technology, a cornerstone of astrophotonics. A/Prof Leon-Saval has been a member of technical program and management committees on more than 10 international conferences. He is a Senior Member of the Optica (OSA), and elected Council Member of the Australian Optical Society (AOS). He was the 2019 recipient of the AOS John Love Award, that recognises innovations and technical advances in the field of optics.

ACKNOWLEDGEMENTS

We would like to thank our colleagues at Australian Astronomical Optics and Physics and Astronomy at Macquarie University, and within the Sydney Astrophotonic Instrumentation Laboratory at The University of Sydney, for countless discussions over many years which have helped to form the ideas presented in this book. We are particularly indebted to Martin Ams, Chris Betters, Julia Bryant, Mark Casali, Nick Cvetojevic, Peter Gillingham, Michael Goodwin, Simon Gross, Anthony Horton, Nem Jovanovic, Kyler Kuehn, Steve Kuhlmann, Jon Lawrence, Pufan Liu, Barnaby Norris, Gordon Robertson, Will Saunders, Chris Schwab, Hal Spinka, Peter Tuthill, Jessica Zheng, amongst many others. We are eternally grateful for the patience of our families while writing this book, which without their support would not have been possible.

CONTENTS

PART I

CONCEPTS OF ASTRONOMICAL INSTRUMENTATION AND PHOTONICS

CHAPTER 1

INTRODUCTION

Innovation in astronomical instrumentation is essential to the progress of astronomy and astrophysics. All physical theories must be tested through experiment, and in astronomy observations ordinarily take the place of laboratory experiments. Advances in our understanding of physical theories require testing those theories in unexplored regions of parameter space; advances in our understanding of astrophysical theories require observations in unobserved regions of parameter space. In short, to advance our knowledge of astronomy and astrophysics requires new and better observations, and this usually requires new and better instruments.

Observations may be bettered in various ways. Building larger telescopes is an obvious brute-force method to collect more light and thereby increase the sensitivity of observations. The history of observational astronomy shows that the diameter of the largest telescopes[1] in the world have doubled approximately every 54 years since their invention, highlighting the importance of deeper and more sensitive observations for the development of astronomy, see Figure 1.1 (see also Bely 2003).

Equally important to the building of larger telescopes is the development of better instruments. Indeed, the innovation of new instruments offers more opportunities to improve observations than simply collecting more light. The sensitivity of observations can be improved by increasing

[1] This monograph is concerned only with 'optical' telescopes and instrumentation, i.e. those working in the wavelengths ~300–5,000 nm which use technology similar or identical to that for observations at visible wavelengths. Unless otherwise stated the terms telescope, instrument, etc. are restricted to mean optical telescope, instrument etc.

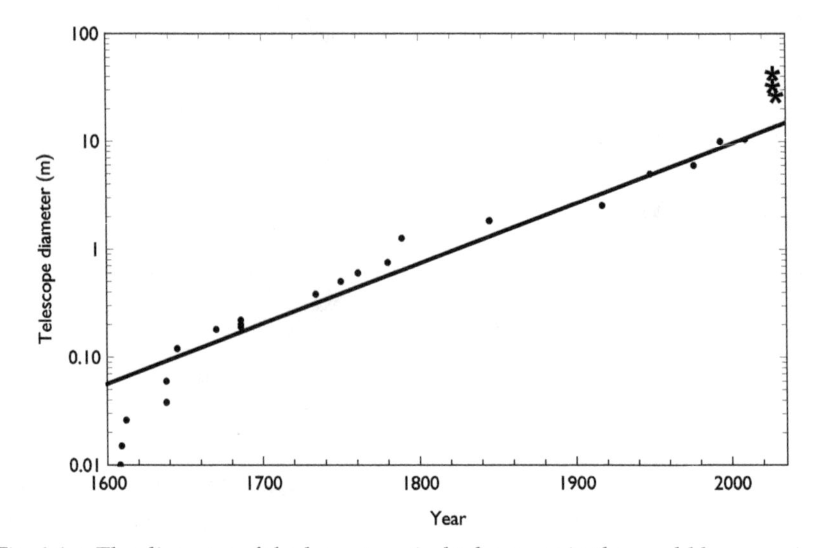

Fig. 1.1. The diameter of the largest optical telescopes in the world by year since their invention (black points). The diameter has doubled approximately every 54 years. Starred points indicate the planned sizes for the next generation of extremely large telescopes: the 24.5 m Giant Magellan Telescope in 2029, the 30 m Telescope in 2027, and the Extremely Large Telescope in 2027.

instrument throughput or efficiency; the precision of observations can be increased by improving instrument stability or calibration methods; the spectral resolving power can be tailored to specific requirements; the spatial resolution can be improved through adaptive optics; the wavelength range of observations can be expanded through the use of new detectors and materials, and so on. Moreover, these innovations can be implemented much more cheaply than building larger telescopes, and a single telescope can host a suite of different instruments for different purposes, which may be updated as technology develops.

Furthermore, genuinely innovative instrumentation can lead to very significant advances. Historical examples include the development of astronomical photography, and also its subsequent replacement with charge-coupled devices (CCDs); the development of spectroscopy, the development of polarimetry; the development of interferometry; the extension of observations to wavelengths other than visible; the correction of atmospheric turbulence. These developments have led to discoveries too numerous to list, and indeed have given rise to whole fields of research which comprise all of modern optical astronomy,[2] all of which were brought about by new instrumental techniques.

This monograph examines the rise of a new field of astronomical instrumentation, *viz.* astrophotonics. Astrophotonics is the application of photonics to astronomical instrumentation. This rapidly developing field has reached a stage of development where many prototype devices are now being tested on sky, and the first fully-fledged instruments incorporating photonic devices are now being used for observations. The field is thus transitioning from one of instrumental research and development to mainstream observational astrophysics. This book is intended to communicate the current status, potential, and future possibilities of the field to the wider astronomical, optics and photonics communities during this period of transition.

The rise of astrophotonics marks a significant departure from the traditional approach to astronomical instrumentation. Astronomical instruments typically employ bulk optics, such as lenses, mirrors, diffraction gratings and filters to manipulate beams of light that propagate in free-space between the optics. Astrophotonic instruments replace such bulk optics with devices embedded in optical fibres and waveguides, such that the manipulation of light takes place within the waveguides themselves. This allows improvements in efficiency, functionality and scalability of astronomical instruments.

Astrophotonics will become increasingly important in the future of astronomical instrumentation, as we enter the era of ELTs — extremely large telescopes with diameters >20 m. Such rapid increase in telescope size requires new ways to think about instrumentation. The full advantage of larger telescopes is realised with adaptive optics (AO), which corrects for the blurring of images induced by atmospheric turbulence, and allows telescopes to operate at or close to their diffraction limit, i.e. with an angular resolution of λ/D, where λ is the wavelength and D is the telescope diameter. Thus, AO provides a D^4 gain in signal-to-noise ratio for background limited observations, as the light gathering power of the telescope is increased by a factor D^2 due to the larger collecting area, and the background is reduced by a factor D^2 due to the smaller angular resolution. For this reason all modern large telescopes are equipped with AO systems.

[2]And those areas of observational astronomy which are not covered here, e.g. radio, X-ray, etc. were brought about by equally, if not more, innovative instrumentation.

Astrophotonics is ideally suited to exploit the advantages offered by AO. Many astrophotonic devices rely on single-mode waveguides (see Chapter 4 for a discussion of waveguide modes), and AO provides a nearly diffraction limited beam that can be coupled efficiently into a single-mode waveguide (as will be discussed in Chapter 6). As we enter the era of ELTs astrophotonics will become more powerful and important.

Astrophotonics is not so much a new technique, as a whole new approach to developing instruments. That is to say, astrophotonics is not directed with a single aim, but rather promises new ways of addressing almost all aspects of optical astronomical instrumentation, as well as introducing novel capabilities.

Before discussing astrophotonics proper, we must therefore define what we mean by photonics. In the remainder of this chapter, we describe the history and development of photonics, and its application to astronomy, and describe the scope of the book.

1.1 Photonics

The field of photonics has its origins in the development of lasers, LEDs, and fibre-optics for communications. The term photonics is analogous to the term electronics; whereas electronics refers to the control of the flow of electric charge, photonics refers to the control of the flow of photons, be that through free-space or through materials. One early definition, from Pierre Aigrain in 1967, reads,

> Photonics is the science of the harnessing of light. Photonics encompasses the generation of light, the detection of light, the management of light through guidance, manipulation, and amplification, and most importantly, its utilisation for the benefit of mankind.

Thus, photonics is very broad field; it covers almost any discipline concerning photons, be that classical optics, or quantum optics or anything in-between. However, the emphasis is on *harnessing* light, i.e. the processing of optical information via the manipulation of light, and especially the technology which enables this.

It is perhaps more expedient to give an explanation of photonics through illustrative examples, rather than a formal definition. An excellent example is telecommunications. Consider a system in which information is encoded into pulses of light. Multiple signals, e.g. using different wavelengths, polarisations or spatial modes, can be combined into a single waveguide using a multiplexer, and transmitted vast distances along an optical fibre. At intervals the pulses may be amplified to boost the signal. Various signals can be routed along different paths of a network using switches. The different signals can be separated using a demultiplexer. Finally, the signals can be detected and converted to an electronic signal. With the exception of the final stage, photonics enables all these processes to be carried out without ever converting the optical signal into an electronic signal. An explanation of the procedures by which such processes are achieved must await the concepts developed throughout Part I. This example is but a tiny snapshot of a broad and burgeoning field, and we could have equally well chosen examples from computing, sensing, medicine, biology, defence, data storage, or fundamental physics.

1.2 Astrophotonics

Astrophotonics is the implementation of photonic devices in astronomical instruments. For the purposes of this book we adopt a more pragmatic definition than that given above, focussing on those photonic technologies of greatest interest for astronomical instrumentation. Within this narrower scope we hereafter regard photonics as the implementation of optical devices within waveguides, be they fibres, planar waveguides, 3D waveguides or any other such. Astrophotonics is therefore concerned with astronomical instruments which incorporate waveguides.

Most prosaically this means astronomical instruments which use fibre optics to guide light from one part of the optical train to another, e.g. from the telescope focal plane to the entrance slit of a spectrograph. More interestingly though, and the real heart of astrophotonics, this means waveguides which include some other optical function, e.g. filters, switches, wavelength division multiplexing (WDM) etc. This is where the excitement and promise of astrophotonics lies, because incorporating complex optical functions within a waveguide offers some truly novel possibilities for astronomical instruments. For example, optical elements may be miniaturised and made modular reducing the size and cost of instruments and allowing extremely flexible functionality. Similarly, some aspects of the performance of instruments may be improved with photonic techniques. Perhaps most exciting is the possibility of new optical functions, which are impossible, or at least extremely difficult to achieve with classical optics, in the same way that photonics is providing new behaviours which are impossible to replicate with classical optics.

A further distinction between photonics and classical optics is the manner in which calculations are made. In many instances in classical and astronomical optics, it is sufficient to use the approximations of geometric optics, and wave optics; and indeed the optical design of most astronomical instruments are carried out using these methods. However, these approximations are not sufficient to describe many properties of photonic devices, for example the evanescent coupling of light between two waveguides. Instead one must use electromagnetic optics, in which the modes of propagation of the electromagnetic field within the waveguides are calculated, or even a full quantum mechanical treatment of individual photons; see the essay in the preface. This distinction between the treatment of optics also leads to differences in the language of the two traditional astronomical optics and photonics.

The scope of astrophotonics may be appreciated from an examination of Part II, in which we introduce astrophotonic devices incorporated into instruments that are already built, or currently are under development. Such developments incorporate the astronomical fields of multi-object spectroscopy, integral field spectroscopy, interferometry, and calibration; using photonic devices for light transport, beam combination, spatial filtering, spectral filtering, optical path length matching, multimode to single-mode conversion, frequency comb generation and WDM. The quiver of techniques used to accomplish these functions is even more numerous (Figure 1.2).

Most progress so far has been made in the fields of spectroscopy and interferometry. However, if one looks to current developments in photonics, which have not yet been exploited or investigated in an astronomical context (Part III), then astrophotonics may be applied to almost all aspects of astronomical instrumentation.

1.3 Scope of the Book

In this book, we aim to provide an introduction to the principles of photonics for the astronomer, and likewise an introduction to the astronomical applications and requirements of astrophotonic instruments for the photonicist. To keep the text accessible to both audiences we aim to emphasise the underlying principles behind the physics of the photonic devices discussed, and likewise behind the astronomical requirements, such that an intuitive and physical understanding of astrophotonics may be obtained. We assume a basic undergraduate-level background

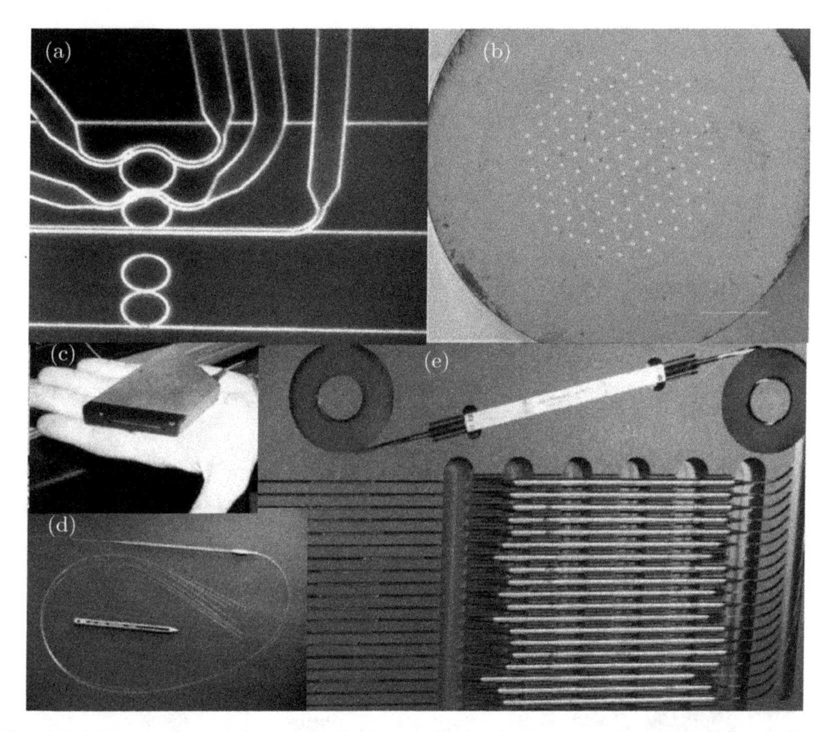

Fig. 1.2. Examples of some of the astrophotonic devices described throughout this book: (a) microring resonators; (b) multicore fibres; (c) arrayed waveguide gratings; (d) photonic lanterns; (e) athermally packaged fibre Bragg gratings and photonic lanterns.

knowledge of optics. Where appropriate we give derivations, but usually only for simplified or particularly important results. For more technical and complicated results, we prioritise clarity and accessibility over formality; if necessary we omit lengthy derivations and quote results directly. It is hoped that these more technical results can be understood in terms of the simpler heuristic models preceding them. More rigorous treatments and derivations may be found in the references provided.

The book is written in three parts, the first developing the background in astronomical instrumentation and photonics necessary to understand astrophotonics. The second part describes developments in astrophotonic instruments themselves, and the third part discusses the future of astrophotonics. Figure 1.3 sketches the general outline of the book for easy reference. Thus, after an introduction to the field in Chapter 1, we give a broad overview of the requirements of astronomical instruments in Chapters 2 and 3. These chapters are intended as a basis for the following

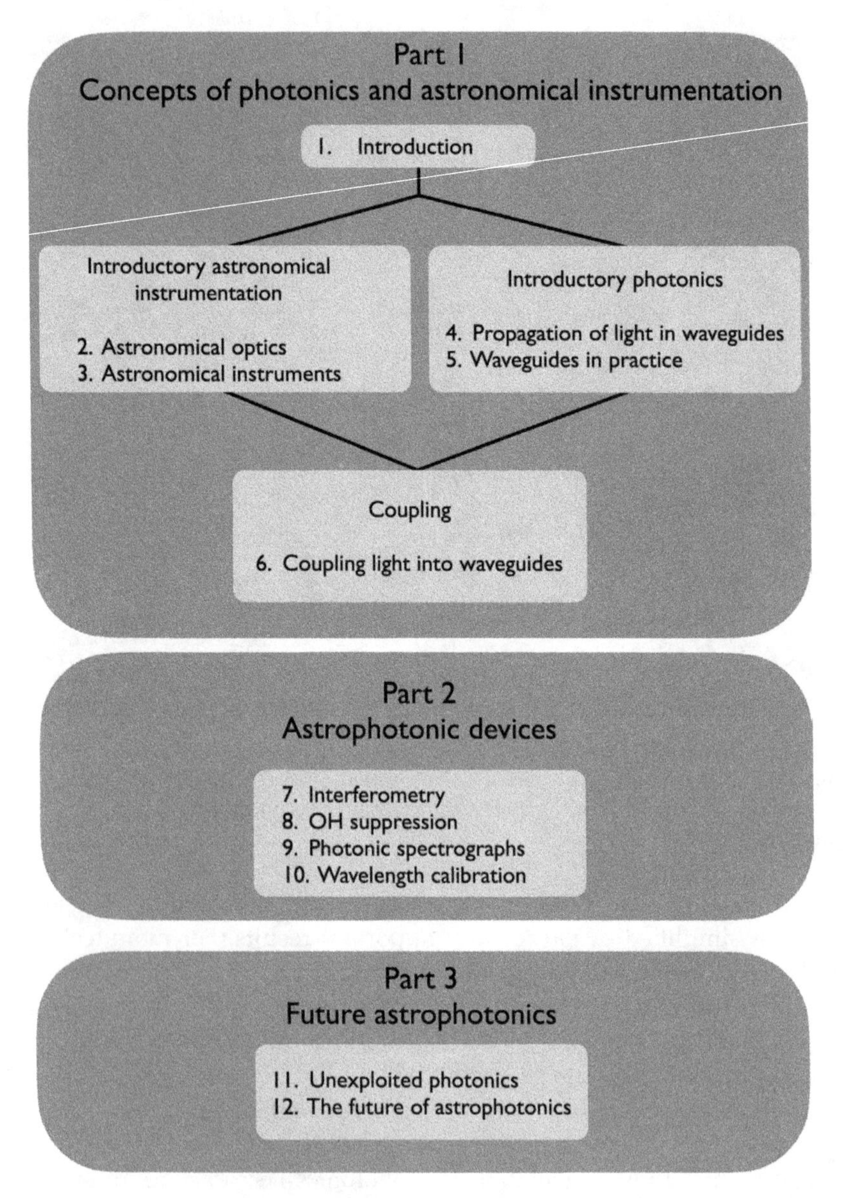

Fig. 1.3. A sketch of the outline of the book.

Table 1.1. Timeline of selected milestones in astrophotonics.

Year	Milestone	Comments
1980	Fibre multi-object spectroscopy	Hill *et al.* (1980)
1992	FLUOR	Fibre interferometric beam combiner Coudé du Foresto and Ridgway (1992)
1997	Integrated optics beam combiner	Kern *et al.* (1997)
2004	Fibre Bragg grating OH filters	Bland–Hawthorn *et al.* (2004)
2005	Photonic lanterns	Leon-Saval *et al.* (2005)
2009	Special issue on astrophotonics	*Optics Express* 12 papers Bland–Hawthorn *et al.* (2009)
2011	PIONIER	Photonic interferometric beam combiner Le Bouquin *et al.* (2011)
2012	GNOSIS	FBG OH suppression instrument Ellis *et al.* (2012a)
2012	Arrayed waveguide gratings on sky test	Cvetojevic *et al.* (2012a)
2012	Dragonfly	Photonic pupil remapping interferometry Jovanovic *et al.* (2012b)
2015	VAMPIRES	Photonic pupil remapping interferometry Norris *et al.* (2015)
2017	GRAVITY	Photonic optical interferometry Gravity Collaboration *et al.* (2017)
2017	Special issue on astrophotonics	*Optics Express* Bryant *et al.* (2017)
2019	Review of integrated photonics	Norris and Bland–Hawthorn (2019)
2019	Astrophotonics white paper USA Decadal Survey on Astronomy and Astrophysics	Gatkine *et al.* (2017)
2020	PRAXIS	FBG OH suppression instrument Ellis *et al.* (2020)
2021	Major review of astrophotonics	Minardi *et al.* (2021)
2021	Double issue special feature on astrophotonics	*JOSA B & Applied Optics* Dinkelaker *et al.* (2021a,b)

discussion of the application of photonics within astronomy; Chapter 2 gives the basic principles of astronomical optics, and Chapter 3 describes various classes of astronomical instruments and their requirements. These chapters are aimed primarily at photonicists coming to astronomy for the first time.

Following this we review the principles of waveguides in Chapters 4 and 5. Chapter 4 describes the propagation of light in waveguides in general, and introduces important concepts. This is necessary background for all that follows, and provides a framework in which the behaviour of all the photonic devices discussed in this paper may be understood. Chapter 5 then discusses the properties of real types of waveguides such as are used in astrophotonics. These chapters are aimed primarily at astronomers coming to photonics for the first time.

Next, Chapter 6 brings together the previous chapters, discussing the crucial problem of coupling light from telescopes into photonic devices, from photonic devices back into traditional astronomical instruments, and coupling between waveguides. This is a critical issue for the astronomical implementation of photonic devices.

After this introductory background, Part II reviews various types of astrophotonic instruments currently in use, under development, or being planned. We hope that these chapters provide a useful review of the state of astrophotonics today. We do not give a history of the development of these instruments; the field is still nascent and such a history would be premature; however, we provide key references for all the developments discussed herein. Rather, we again wish to provide the principles of these instruments, as well as their capabilities and performance. Some key review papers and milestones in astrophotonics are given in Table 1.1, which gives some small indication of the progress and history of the field, focussing on those technologies which have been successfully tested at major observatories.

In the final chapters of Part III, we look to future developments. Chapter 11 discusses existing photonic technologies with potential astronomical applications, but which have not been exploited. Finally, Chapter 12 discusses possible future directions for astrophotonic applications based on developments yet to take place, including multimode astrophotonics, wholly photonic astronomical instruments, and quantum photonics. Such instruments would fully realise the potential benefit of photonics for astronomy, enabling capabilities and modularity not otherwise possible.

CHAPTER 2

ASTRONOMICAL OPTICS

Astrophotonics brings a new approach to building astronomical instruments in which photonic components replace the traditional bulk optics components such as lenses and mirrors. Nevertheless, traditional optics remain an indispensable part of astrophotonics. In the first place, all astrophotonic components are coupled to a telescope, which is itself an optical component. Moreover, to date and for the immediate future, all astrophotonic instruments are comprised of both photonic components and bulk optics; the photonic components are incorporated within a traditional optical instrument. We examine the potential for fully photonic instruments in Chapter 12.

In this chapter, we introduce the basic principles of astronomical optics, which are requisite for an understanding of astrophotonics, focussing on general principles. In the next chapter, we will look at specific types of astronomical instruments and their requirements.

2.1 The Seeing

Throughout this book, we are concerned mainly with instruments for ground-based telescopes. Therefore we are concerned with observations made through the Earth's atmosphere. The atmosphere has a significant detrimental effect on observations, first because it deteriorates the image quality, and secondly because it introduces a background light which must be subtracted from observations. The latter problem will be discussed later (Chapter 8); here we discuss the former.

Seeing is the distortion of the incoming wavefronts by turbulence in the Earth's atmosphere. Turbulence mixes layers of air which have different temperature and therefore different refractive indices. This causes the incoming light to be both blurred and displaced; the magnitude of this effect is characterised by the seeing. The seeing can be described by Fried's parameter, r_0, which describes the diameter of a bundle of rays which are initially parallel and in phase, and remain so after passing through the atmosphere and arriving at the telescope aperture. For a telescope with an diameter, $D \gg r_0$, the full width at half-maximum (FWHM), of the image is given by (Bely 2003)

$$\mathrm{FWHM} = 0.98\frac{\lambda}{r_0} > \frac{\lambda}{D}, \tag{2.1}$$

where λ is the wavelength, and the inequality on the right hand side gives the diffraction limit of a telescope of diameter D. Thus, large telescopes are not diffraction limited, unless the seeing distorted wavefront is corrected with adaptive optics (Section 2.5), or unless adopting 'lucky imaging' whereby extremely short exposures are taken and culled to select only those exposures which happened to have unusually good seeing (Fried 1978). Furthermore, note that $r_0 \propto \lambda^{\frac{6}{5}}$, and therefore

$$\mathrm{FWHM} \propto \lambda^{-\frac{1}{5}}, \tag{2.2}$$

and is thus better at longer wavelengths. Typical seeing at a good observing site is $\sim\frac{2}{3}$ arcsec in the visible and $\sim\frac{1}{2}$ arcsec in the near-infrared.

The intensity profile of a point source such as a star in natural seeing conditions can be well modelled by a **Moffat function**,

$$I(r) = \left(1 + \left(\frac{r}{\alpha}\right)^2\right)^{-\beta}, \tag{2.3}$$

where α and β are free parameters. For turbulence induced seeing $\beta \approx 4.765$. The α parameter is related to the FWHM of the profile by,

$$\alpha = \frac{\mathrm{FWHM}}{2\sqrt{2^{\frac{1}{\beta}} - 1}}. \tag{2.4}$$

Figure 2.1 compares a Moffat function to a Gaussian of the same FWHM; the Moffat function has stronger wings, which can be most clearly seen in the logarithmic plot.

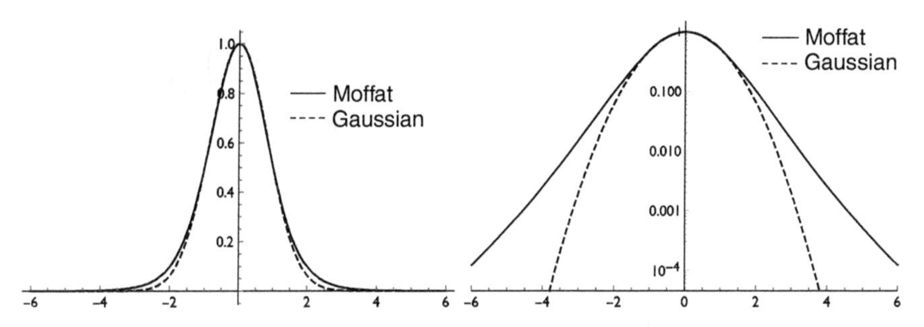

Fig. 2.1. Comparison of Moffat and Gaussian functions, for profiles with the same FWHM, shown on a linear scale (left) and a logarithmic scale (right). The Moffat has more light in the wings of the function, as can be seen clearly in the logarithmic plot.

2.2 Astronomical Telescopes

The telescope is the basis of observational astronomy. Since almost all modern optical astronomy makes use of large reflectors we will restrict our discussion to this class of telescope. In its simplest form the telescope consists of a single primary mirror which collects and focusses light. Before discussing the design of telescopes, we note four important points which inform much of the following discussion, which are illustrated in Figure 2.2.

(1) All astronomical sources are at such vast distances that, in terms of the optics, we can consider them to be at an infinite distance, such that the light rays arriving at the top of the atmosphere are collimated with a flat wavefront.
(2) Atmospheric turbulence distorts the wavefront so that the light arriving at the telescope is not coherent — see Section 2.1.
(3) Because the light incident on the primary mirror is very nearly collimated (within the limits of the seeing), a single mirror (or lens) is sufficient to form an image.
(4) All astronomical sources are large, and the light from each object fills the entire telescope mirror.

2.2.1 *Telescope foci*

The key parameters of a telescope are the diameter of the primary mirror, and the associated collecting area, and the focal length of the telescope.

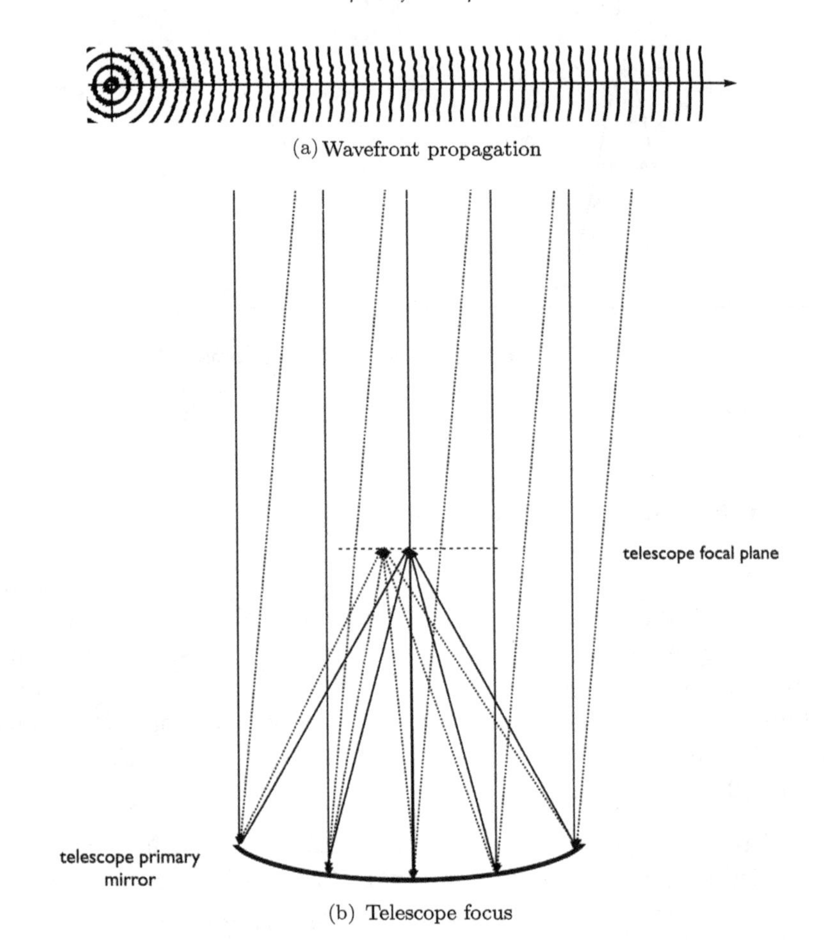

Fig. 2.2. (a) The light from an astronomical point source is initially incoherent, with a random wavefront, but, due to the very large propagation distance, becomes flat by the time it arrives at the top of the earth's atmosphere, and therefore is also collimated. (b) A single mirror is sufficient to form an image, since the incident light is nearly collimated, and the light from each source fills the entire telescope mirror.

Both the collecting area and focal length depend on the configuration of the telescope, which for many telescopes is changeable. Figure 2.3 illustrates the most common foci used in optical astronomy. The actual focal lengths depend on the combinations of mirrors. For instance a single telescope may have a choice of secondary mirrors to give different focal lengths for Cassegrain instruments.

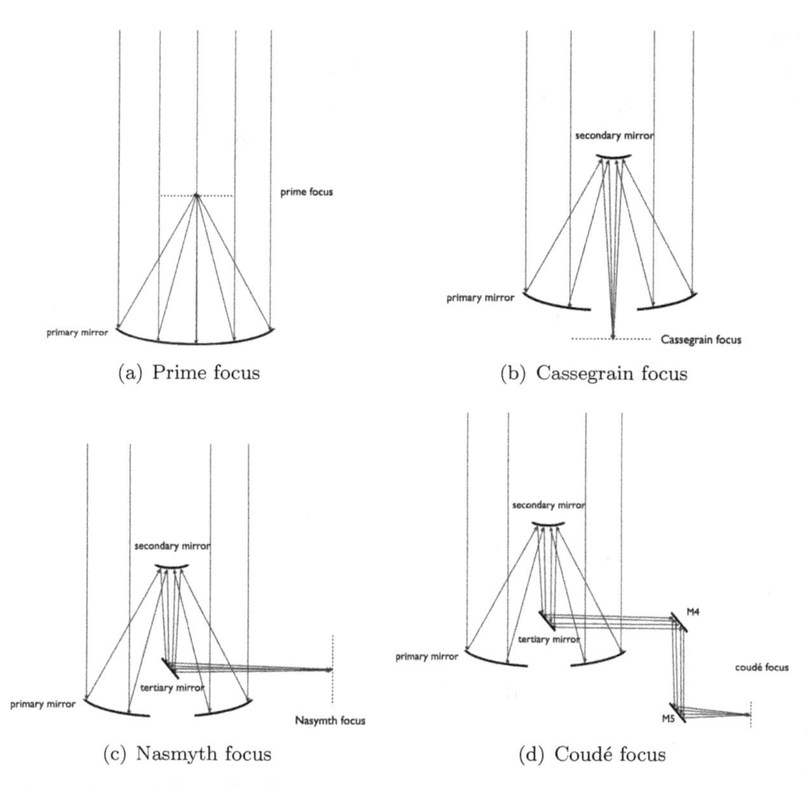

(a) Prime focus

(b) Cassegrain focus

(c) Nasmyth focus

(d) Coudé focus

Fig. 2.3. Typical foci of reflecting telescopes at professional observatories. The same telescope can use several foci by means of changing the 'top end' of the telescope to switch between prime focus instruments and secondary mirrors. Different secondary mirrors can be employed to change the focal length of the telescope, and hence the speed of the focus. The Nasmyth and coudé foci are similar and both require a tertiary mirror, but the Nasmyth is used on alt-azimuth telescopes, and directs the beam through the alt-bearing onto a stable gravity invariant platform. The coudé focus is used mainly on equatorial telescopes, and usually consists of several mirrors to direct the beam to a separate room via the declination axis.

2.2.2 *Collecting area*

The amount of light that a telescope can collect is one of its most important properties. Quite simply, a telescope with a larger mirror can collect more light from any given object, and this scales with the area. The exposure time necessary to reach a given signal-to-noise ratio (see Section 3.1.1) scales as $1/A$, so a telescope with twice the diameter can observe ≈ 4 times faster. This is only approximate since often part of the primary mirror is obscured by the secondary mirror or instruments, as in Figure 2.3.

Note well, that since the light from a star or other astronomical body fills the entire mirror, a central obstruction does not reduce the field-of-view (FOV) of the telescope; the FOV is contiguous. The drive to observe fainter objects has driven the increase in telescope diameter ever since the time of Galileo, see Figure 1.1.

2.2.3 *Focal ratio*

The **focal ratio** (or f-ratio) of the telescope is an important parameter in telescope design, and is simply the focal length divided by the diameter of the primary mirror,

$$F = \frac{f}{D}. \tag{2.5}$$

The focal ratio is often referred to as the **speed**, with small focal ratios being fast, and high focal ratios being slow. These terms, which have their origins in photography, are still in common use in astronomical optics. A smaller focal ratio has a larger solid angle, and therefore collects more light per unit area at the focal plane. Thus, for a given detector a shorter exposure time can be used; i.e. the exposure is *faster*, whereas for a larger focal ratio the exposure would be *slower*.

2.2.4 *Plate-scale*

The focal ratio determines the **plate-scale** of the telescope, p, which is the number of arcseconds per mm (or equivalent units) at the focal plane. Consider Figure 2.4 in which a star subtends an angle θ at telescope with primary mirror diameter D and focal length f, and forms an image of size y at the focal plane. The plate-scale is given by

$$p = \frac{\theta}{y}. \tag{2.6}$$

Since $y = f \tan\theta$, for small angles this becomes

$$p \approx \frac{1}{f} = \frac{1}{FD}. \tag{2.7}$$

Or equivalently

$$p \approx \frac{\alpha}{D}, \tag{2.8}$$

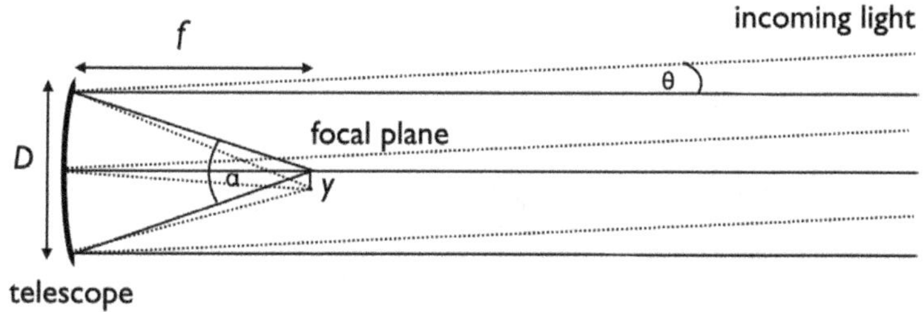

Fig. 2.4. Sketch of a simple telescope to define the plate-scale.

i.e. the ratio of the angle subtended by the primary mirror at the focal plane, α, to the diameter, D, of the mirror. Therefore, in arcseconds per unit length this simply becomes,

$$p \approx \frac{206{,}265}{FD}, \tag{2.9}$$

since there are 206,265 arcseconds in one radian.

Ideally one would like the plate-scale to be matched to the requirements of the instrument. For example, for most imagers this means that the angular resolution per pixel should be sufficient to properly sample the point spread function (PSF). The size of the PSF is either determined by the turbulence of the atmosphere (seeing), the adaptive optics correction of atmospheric turbulence, or the diffraction limit of the telescope (Section 2.1). In any case, the PSF should be sampled by ≥2 pixels in diameter, to satisfy Nyquist sampling theory, see Figure 2.5.

Example 2.1: What is the optimal plate-scale and focal-ratio to image stars of angular diameter 1.5 arcsec using a telescope with a 4 m diameter mirror equipped with a detector with 15 μm pixels?
We require the angular diameter of star to be sampled by two pixels, so the required plate-scale is,

$$p = \frac{1.5}{2 \times 15 \times 10^{-3}} = 50 \text{ arcsec mm}^{-1}.$$

The necessary focal-ratio is given by,

$$p = \frac{206{,}265}{F \times 4{,}000}$$
$$\implies F \approx 1.03.$$

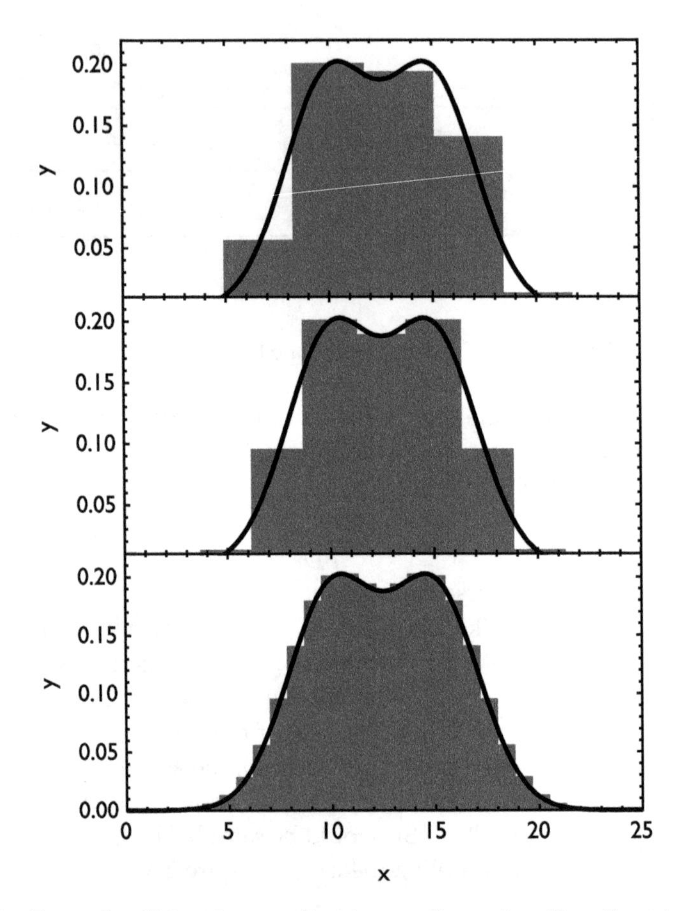

Fig. 2.5. Example of Nyquist sampling in one dimension. Two Gaussians separated by the FWHM are just resolved (black lines). When sampled by the FWHM/2 (grey histogram, middle panel), the resulting data can just recover the presence of two individual Gaussians. When under-sampled at FWHM/1.5 (grey histogram, top plot) the two Gaussians can no longer be distinguished. Over sampling by FWHM/6 (grey histogram, bottom plot) divides the signal into an unnecessarily high number of pixels.

In practice, the choice of the focal ratio of the primary mirror is constrained by other considerations. Shorter focal lengths (i.e. smaller focal ratios) lead to more compact telescope designs which are cheaper and mechanically stiffer, leading to better image quality, and requiring smaller secondary mirrors. However, the fabrication tolerances on the mirror are more stringent, as is the alignment tolerance of the secondary mirror

(Bely 2003). Nevertheless, it is now usual for telescope primaries to be made as fast as possible.

For specific science cases other considerations may override the matching of the plate-scale to the Nyquist sampling. For instance, if sensitivity is a priority then the image may be deliberately under-sampled in order to increase signal-to-noise per element at the expense of spatial resolution, which will not be able to properly recover the images, or be able to distinguish between point sources and cosmic rays. On the other hand, if high spatial resolution is a priority, then the instrument may oversample the image at the expense of decreased signal-to-noise, due to increased detector noise.

2.3 Re-Imagers

We have seen that it is beneficial to match the plate-scale to the instrument, e.g. to match the pixel size of a detector, or the slit width or fibre diameter of a spectrograph. However, in practice the two are not usually commensurate. The telescope focal ratio may have been selected for various reasons, e.g. an older telescope may have been designed to match the speed of photographic plate emulsions, or a modern telescope may have been designed to be as fast as possible for mechanical stiffness. Likewise, the pixel size, or fibre diameter may depend on the available technology. Therefore, it is usually necessary to magnify (or de-magnify) the telescope beam.

In principle this could be achieved with a single lens, as in Figure 2.6(a). However, although simple, this **focal reducer** arrangement has some disadvantages. For example, inserting a filter before the detector will shift the location of the focal plane, and thus care must be taken that all filters have exactly the same thickness and refractive index. Moreover, for certain types of filter, such as interference filters, the filter bandpass shifts as a function of incident angle, and thus a converging beam will have the effect of blurring and broadening the filter bandpass. Furthermore, tilting the filter to avoid unwanted reflections at the detector (known as ghosts), will introduce lateral astigmatism which must then be corrected by the optics.

Due to such considerations it is often preferable to use a collimator-camera arrangement to change focal ratio, such as shown in Figure 2.6(b), in which the beam is first collimated, and then re-focussed at the desired focal ratio. Now any filters can be placed in the collimated beam without

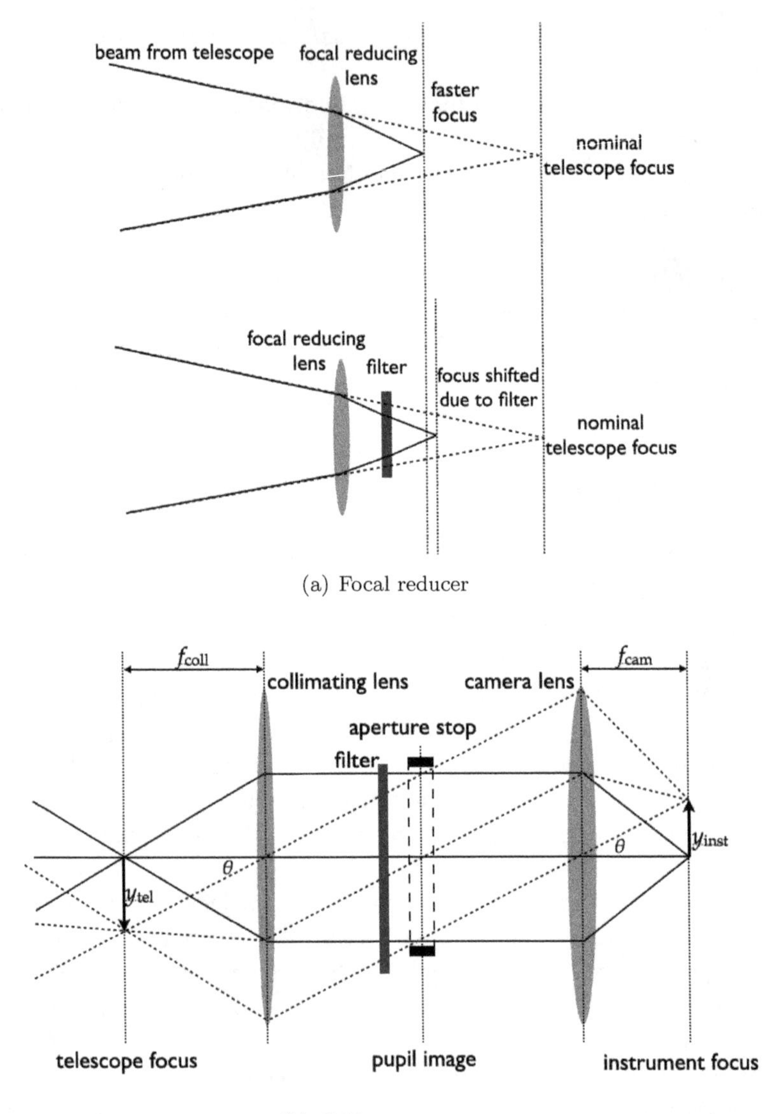

Fig. 2.6. (a) A single lens can change the focal ratio of the telescope beam to match a required plate-scale. However, any elements placed within the converging beam will cause a shift in the location of the focal plane. (b) A collimator-camera arrangement can also change the focal ratio, but now optics such as filters or diffraction gratings can be placed within the collimated beam without altering the location of the focal plane.

affecting the location of the focal plane, and likewise stops and baffles can be placed at the intermediate pupil image to prevent stray-light from reaching the detector. Furthermore, the same basic collimator-camera arrangement can be used as the basis for a spectrograph by inserting a dispersing element, such as a diffraction grating, into the collimated beam.

The change in focal ratio due to a collimator-camera, or a focal reducer, magnifies the image. For example, in Figure 2.6(b), the image at the telescope focus has a physical size of

$$y_{\text{tel}} = f_{\text{coll}} \tan\theta. \tag{2.10}$$

Likewise, at the instrument focal plane the image has a physical size of

$$y_{\text{inst}} = f_{\text{cam}} \tan\theta. \tag{2.11}$$

Thus, the magnification is

$$M = \frac{y_{\text{inst}}}{y_{\text{tel}}} = \frac{f_{\text{cam}}}{f_{\text{coll}}}. \tag{2.12}$$

Note that

$$F_{\text{tel}} = \frac{f_{\text{tel}}}{D_{\text{tel}}} = \frac{f_{\text{coll}}}{D_{\text{coll}}}, \tag{2.13}$$

where D_{coll} is the diameter of the collimating lens. Therefore, if $D_{\text{cam}} = D_{\text{coll}}$, then

$$M = \frac{F_{\text{cam}}}{F_{\text{tel}}}. \tag{2.14}$$

So the magnification is simply the ratio of the final focal ratio to the telescope focal ratio. The final plate-scale is

$$p = \frac{206{,}265}{F_{\text{cam}} D_{\text{tel}}} = \frac{206{,}265}{M F_{\text{tel}} D_{\text{tel}}} = \frac{p_{\text{tel}}}{M}. \tag{2.15}$$

Thus, if $M > 1$ the image is bigger at the instrument focus. This corresponds to a smaller overall FOV, and a smaller plate-scale. If $M < 1$, then the image is smaller at the instrument focus the FOV is larger, and the plate-scale is larger.

Example 2.2: Consider a 4 m diameter telescope, with a focal ratio of $f/8$. It is required to observe galaxies of 2″ angular size with a fibre of core size 100 μm. Calculate the necessary magnification to achieve this. The nominal plate-scale of the telescope is

$$p_{\mathrm{tel}} = \frac{206{,}265}{8 \times 4{,}000} = 6.45 \text{ arcsec mm}^{-1}.$$

At the nominal telescope focus a 2″ galaxy would therefore cover 310 μm. The desired plate-scale is

$$p_{\mathrm{inst}} = \frac{2}{0.1} = 20 \text{ arcsec mm}^{-1}.$$

Therefore, the necessary magnification is

$$M = \frac{p_{\mathrm{tel}}}{p_{\mathrm{inst}}} = \frac{6.45}{20} = 0.32,$$

or equivalently

$$M = \frac{y_{\mathrm{inst}}}{y_{\mathrm{tel}}} = \frac{100}{310} = 0.32.$$

2.4 Conservation of Étendue

Consider the simple imaging system shown in Figure 2.7. The light from the star being observed subtends a solid angle Ω_1 and is collected by the telescope aperture which has an area A_1, which is in a medium of refractive index n. The product $n^2 A_1 \Omega_1$ is known as **étendue**. The étendue is generally conserved throughout the entire optical train, such that if an element magnifies the beam, causing the solid angle to lessen, the associated area will increase. This principle is called the **conservation of étendue**.

Similarly, for optical systems which are all on-axis, the **linear étendue**, $n\, r\, \sin\theta$, is also conserved, where r is the radius of the aperture or optic, and θ is the half-angle subtended by it.

If étendue is not conserved throughout the optical train it will have adverse consequences. If the étendue gets smaller then light will be lost, causing a decrease in efficiency. For example, consider if a pupil stop were placed at the pupil image in Figure 2.7 which was smaller than the area of the pupil image. This would decrease the étendue by vignetting the beam, causing a decrease in efficiency.

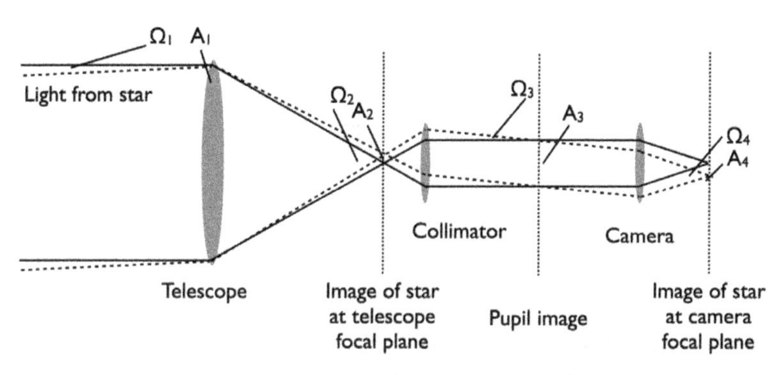

Fig. 2.7. Sketch of a simple imaging system illustrating the conservation of étendue. At each point in the system the product of the square of the refractive index, n^2, solid angle, Ω and the area corresponding area, A, at each component is conserved.

On the other hand, if the étendue of any component is too large then that component will accept light from outside the beam, causing an increase in the background, and a decrease in the signal-to-noise. If the étendue of the beam itself is increased, as it can be by optical fibres (see Section 4.2.5), then all subsequent optical components will require a larger étendue, requiring larger or faster optics, both of which will increase the cost of the instrument. Furthermore, the light will be more spread out, reducing the surface brightness, and therefore the signal-to-noise, e.g. by requiring more pixels for each resolution element.

2.5 Adaptive Optics

Atmospheric turbulence means that the angular resolution of ground-based telescopes is not set by the theoretical diffraction limit of the telescope, but by the seeing (Equation 2.1).

Many astrophotonic devices rely on single-mode waveguides (see Chapter 4 for a discussion of waveguide modes). As will be seen, high coupling efficiency from a telescope into a single-mode fibre can only be achieved if working at the diffraction limit (Section 6.2.2.1). One way to achieve this is in seeing limited conditions is to correct the effects of atmospheric turbulence with **adaptive optics**, commonly abbreviated to **AO**, to reproduce images which are close to diffraction limited.

Adaptive optics is a large field which is beyond the scope of a book on astrophotonics. There are many excellent texts and reviews on this

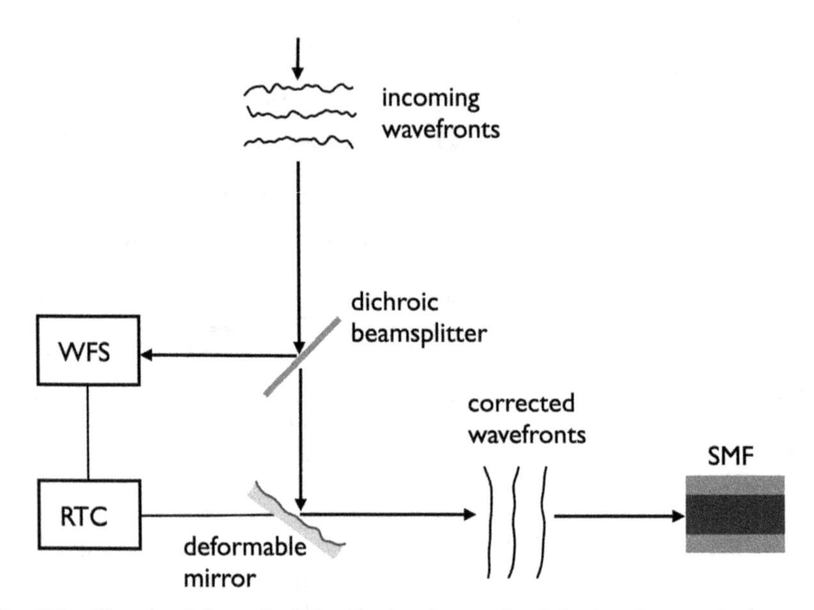

Fig. 2.8. Sketch of the principle of adaptive optics injection into a single-mode fibre, see text for details.

rapidly growing and diverging field; for example Hardy (1998) gives a good overview of the theory of adaptive optics. For our purposes, it suffices to give a brief overview of the principles, as sketched in Figure 2.8. The incoming wavefronts from a point source are distorted by the turbulent atmosphere (Section 2.1), significantly blurring the image. The point source may be either a bright star near (in projection on the sky) to the object being observed, or it may be a 'laser guide star', which is formed by shining a bright laser with a wavelength of ≈ 589 nm into the atmosphere, which excites the Na D transition of Na atoms at about 90 km altitude, which then appear as a bright point source. Using several lasers to form a constellation of point sources can allow the correction of a larger field. Light from the guide star is reflected with a dichroic beam-splitter and sent to a wavefront sensor (WFS). A real time computer (RTC), calculates the distortions in the incoming wavefront, and sends a signal to a deformable mirror. This very thin mirror is then deformed by hundreds or thousands of actuators in such a way as to nearly cancel out the distortions in the wavefront. The light reflected from the deformable mirror is then sent to a science instrument. In a real system there may be several deformable mirrors to correct low or high order aberrations, or different layers of atmospheric turbulence.

2.6 Detectors

A full discussion of astronomical detectors is outside the scope of this book, so here we will give the basic details for the most relevant types of detectors for current astrophotonic developments. For a thorough discussion on astronomical detectors, see McLean (2008).

In this book, we only consider electronic imaging detectors, which can be thought of as a rectangular array of light sensitive pixels constructed from a semiconductor. When a photon of an appropriate wavelength is incident on the detector it excites an electron into the conduction band of the semiconductor via the photoelectric effect, whereupon the electron is trapped in the potential well of the pixel by the detector electronics. The number of electrons accumulated in each pixel throughout an exposure therefore measures the flux of light at that pixel.

The two main types of detector applicable for astrophotonic instruments are **charge-coupled devices (CCDs)** and **complementary metal oxide semiconductor (CMOS)** arrays. We discuss some of their basic properties in the following sub-sections.

2.6.1 *Charge-coupled devices (CCDs) and CMOS detectors*

CCDs are the typical detector of choice for observations from ~350 to 900 nm, offering high quantum efficiency and low detector electronic noise. CCDs are available in large formats compared to other technologies, with arrays up to 10,000 × 10,000 pixels available as commercial 'off-the-shelf' devices.

CCDs use a silicon semiconductor and can be optimised for specific wavelengths. For example, in astronomical applications it is common to illuminate a CCD from the back side (opposite to the electronics), to avoid the absorption due to the electronics on the front side. However, different wavelengths of light will penetrate different distances into the silicon material before releasing photoelectrons, with shorter wavelengths penetrating shorter distances. Therefore, for blue sensitive CCDs, back-illuminated, back-thinned CCDs are common, in which the Si layer has

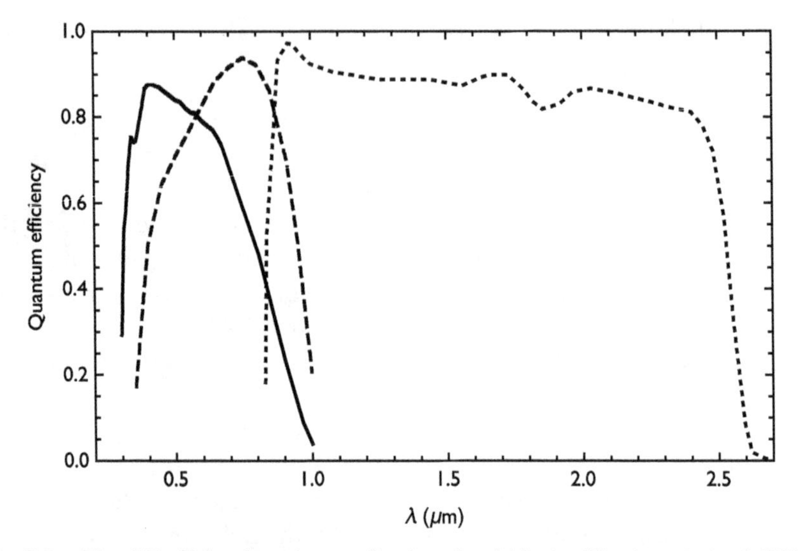

Fig. 2.9. The QE of blue (continuous line) and red (dashed line) optimised CCDs, and of a near infrared HgCdTe CMOS detector (dotted line), as a function of wavelength.

been thinned to 10–20 μm. For higher sensitivity to wavelengths longer than 600 nm back-illumination is still used, but with doped Si to make the light sensitive layer deeper, and therefore more sensitive to longer wavelengths — such devices are known as deep-depletion CCDs. Further optimisation is made by applying appropriate anti-reflection coatings. Figure 2.9 shows the QE as a function of wavelength for a blue and red optimised CCD.

The major limitation of CCDs is the slow read-out time. The charge in each pixel is sequentially transferred to the adjacent pixel along each row, until it reaches the read-out register, when it is shifted along the read-out column, again one pixel at a time, see Figure 2.10. In comparison each pixel of a CMOS detector can be read individually and simultaneously. Slower read-outs better preserve the charge throughout each transfer step. If a single pixel is 'dead', then no charge can be transferred through it, leading to 'dead columns'. For applications in which sensitivity is of primary concern, these read-out overheads are a small price to pay for the excellent QE and noise characteristics of CCDs, especially if the exposure times are much longer than the read-out times. However, CMOS devices are rapidly catching up to the performance of CCDs and it is likely that CMOS will eventually become ubiquitous in astronomy.

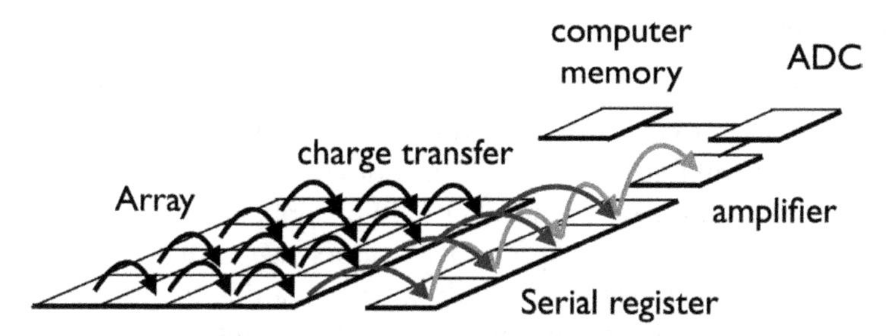

Fig. 2.10. Sketch of the operation of a CCD. The charge in each pixel of the array is shifted one place to the right, with the rightmost pixels transferring to a serial register. The charges in the serial register are then transferred up one pixel at a time into an amplifier, followed by an analog-to-digital converter (ADC), and thence stored in computer memory; then the whole process repeats until the entire array has been read.

2.6.2 *Fast read-out devices*

For applications in which fast read-out times are necessary, e.g. for adaptive optics wavefront sensing, lucky imaging, etc., a different detector will be needed, e.g. frame-transfer electron multiplying CCDs (EMCCDs), avalanche photodiodes (APDs) or CMOS.

In a **frame transfer CCD**, only half the array is exposed. At the end of the exposure, the charges in each pixel are rapidly shifted to the unexposed half of the CCD, whence the signal can be read out through the usual charge transfer technique to the serial register.

In an **EMCCD**, the charge is transferred from the serial read-out register to a multiplication register, which contains hundreds of electrodes. As the photoelectrons are read-out they impact the electrodes generating secondary electrons. In this way the original signal can be multiplied by factors of a thousand or so during the read-out process.

An **APD** is different type of semiconductor imaging array, in which each incident photon induces an 'avalanche' of electrons, through a high internal gain within each pixel. The gain is induced by a strong voltage, which accelerates each photoelectron, which thereby produce secondary electrons in a runaway process. Because the number of electrons per incident photon is highly multiplied, APDs are sensitive to very low light levels. Photon counting devices with frame-rates over 10,000 frames per second are available.

In a CMOS detector each pixel is individually addressed and all pixels can therefore be read-out simultaneously. Read-out speeds of 0.0025 s have been achieved.

2.6.3 *Near-infrared detectors*

At wavelengths >1 μm Si is transparent because the photons do not have sufficient energy to excite an electron across the band-gap, and so CCDs are no longer suitable. Instead a hybrid device, consisting of an infrared sensitive semiconductor, such as HgCdTe or InSb, is bonded to a Si layer containing the electronics. Infrared detectors use CMOS technology in which each pixel can be read out individually, leading to much faster read out times, and avoiding the problem of 'dead columns'. Furthermore, each pixel can be read-out non-destructively, i.e. without erasing the charge in each pixel. This allows one to read out the image several times throughout an exposure, allowing for more sophisticated statistical measurements of the flux of light falling on each pixel, and thereby reducing the effective read out noise, which would otherwise be high.

Because the band-gaps for a NIR detector must necessarily be small to be sensitive to lower energy, longer wavelengths, the dark current can be quite high as it is correspondingly easier to excite thermal electrons into the conduction band. Therefore, NIR arrays must be operated at very low temperatures, ~70 K. Historically NIR CMOS arrays have not been as efficient as CCDs, though with modern devices this is changing, as in Figure 2.9. However, the read-out noise and dark current are still considerably worse than for CCDs.

Since CCDs are generally used for observations <1 μm and hybrid HgCdTe or InSb arrays at >1 μm, the former wavelengths are often referred to as 'visible' or 'optical', despite the fact that wavelengths >700 nm are certainly not actually visible to the human eye. Likewise, the term near-infrared is commonly used in the restricted sense of those wavelength from 1 to 5 μm for which HgCdTe or InSb arrays are used.

CHAPTER 3

ASTRONOMICAL INSTRUMENTS

Having introduced the basic principles of astronomical telescopes and optics in the preceding chapter, we now turn our attention to specific types of astronomical instruments.

The field of astronomical instrumentation is large and diverse. Here we give only a very brief introduction to the most common types of instruments. Rather than presenting a very general overview we focus on specific examples of typical instruments. For a broader survey of astronomical instrumentation a good pedagogical introduction at approximately undergraduate level is Chromey (2010), a more advanced introduction is Kitchin (2021); selected original papers may be found in Livingston (1993) or indeed in any of the proceedings of SPIE conferences in astronomical instrumentation.

In addition to describing the behaviour of astronomical instruments, we also discuss their requirements. The technological requirements of any astronomical instrument are specific to the particular science case it is developed for. However, there are general differences between the requirements of astronomical instruments and those of typical photonic devices, e.g. for telecommunications, which imply a difference in implementation. In practice this means that photonic devices cannot usually be adopted for astronomy without first adapting them.

3.1 Astronomical Measurements

Before discussing specific types of instruments, we give a brief introduction to the different types of measurements made in astronomy.

Within the realm of optical astronomy these can be classified into different types of measurement, *viz.*,

- **Photometry**, the measurement of an object's brightness, i.e. the number or energy of photons produced per second in a given bandpass.
- **Spectrometry**, the measurement of an object's spectrum, i.e. the absolute or relative number of photons produced as a function of wavelength.
- **Imaging**, recording an image of an object or group of objects.
- **Astrometry**, the measurement of the positions of objects on the celestial sphere.
- **Polarimetry**, the measurement of the degree of polarisation of light from an object.
- **Interferometry**, the measurement of the coherence between multiple photons, either from the same source or different sources.
- **Timing**, the timing of transient or periodic events.

These are not necessarily independent measurements, and can be combined, e.g. an image will usually yield photometric and astrometric measurements, and spectrometry can be combined with photometry and imaging, as two simple examples. These concepts will be discussed in more depth in the context of the instruments themselves throughout this chapter.

3.1.1 *Signal-to-noise*

The significance of an astronomical measurement is often expressed as the **signal-to-noise ratio**, which is given by

$$\mathrm{SNR} = \frac{N_\mathrm{S}}{\sqrt{N_\mathrm{S} + N_\mathrm{B}}}, \tag{3.1}$$

where N_S is the number of counts in the signal and N_B is the number of counts in the background, and it is assumed that $N_\mathrm{S} + N_\mathrm{B}$ is large enough that the Poisson noise can be approximated by $\sqrt{N_\mathrm{S} + N_\mathrm{B}}$.

For example, consider a photometric measurement of source which produces S photons s^{-1} m^{-2} μm^{-1}, then

$$N_\mathrm{S} = S\, t\, A\, \eta\, \Delta\lambda, \tag{3.2}$$

where t is the exposure time, A is the collecting area, η is the end-to-end throughput of the instrument from the atmosphere to the detector, and $\Delta\lambda$ is the wavelength range of interest, e.g. the width the passband in which the measurements are made.

The background can be composed of many different elements. Typically important backgrounds are the emission from the Earth's atmosphere (the sky background), the dark current of the detector, D, and the read out noise of the detector, R. For particular instruments other backgrounds may also be important, e.g. thermal emission from the telescope or instrument can be significant for infrared instruments. If the sky background is B photons s^{-1} m^{-2} μm^{-1} arcsec^{-2}, the dark current is D e^- pixel^{-1} s^{-1} and the read-out noise is R e^- pixel^{-1}, then the total background is

$$N_{\rm B} = B\,t\,A\,\eta\,\Delta\lambda\,\Omega + D\,t\,p + R^2\,p, \tag{3.3}$$

where Ω is the area over which the source counts were measured in arcsec^{-2}, and p is the number of pixels on the detector over which the signal is spread.

Example 3.1: Show that if the read-out noise is negligible then $\mathrm{SNR} \propto \sqrt{t}$.

In this case, we have

$$N_{\rm S} = S\,t\,A\,\eta\,\Delta\lambda,$$

$$N_{\rm B} = B\,t\,A\,\eta\,\Delta\lambda\,\Omega + D\,t\,p,$$

and therefore Equation 3.1 becomes

$$\begin{aligned} \mathrm{SNR} &= \frac{S\,t\,A\,\eta\,\Delta\lambda}{\sqrt{S\,t\,A\,\eta\,\Delta\lambda + B\,t\,A\,\eta\,\Delta\lambda\,\Omega + D\,t\,p}}, \\ &= \sqrt{t}\,\frac{S\,A\,\eta\,\Delta\lambda}{\sqrt{S\,A\,\eta\,\Delta\lambda + B\,A\,\eta\,\Delta\lambda\,\Omega + D\,p}}. \end{aligned}$$

$$\Longrightarrow \mathrm{SNR} \propto \sqrt{t}$$

Q.E.D.

Example 3.2: Show that if both the read-out noise and dark current are negligible then $\text{SNR} \propto \sqrt{\eta\, A\, t}$, and thus the exposure time necessary to reach a given SNR scales as $1/A\eta$.

Now, we have

$$N_\text{S} = S\, t\, A\, \eta\, \Delta\lambda,$$

$$N_\text{B} = B\, t\, A\, \eta\, \Delta\lambda\, \Omega,$$

and therefore Equation 3.1 becomes

$$\begin{aligned} \text{SNR} &= \frac{S\, t\, A\, \eta\, \Delta\lambda}{\sqrt{S\, t\, A\, \eta\, \Delta\lambda + B\, t\, A\, \eta\, \Delta\lambda\, \Omega}}, \\ &= \sqrt{t\, A\, \eta}\frac{S\, \Delta\lambda}{\sqrt{S\, \Delta\lambda + B\, \Delta\lambda\, \Omega}}. \\ \implies \text{SNR} &\propto \sqrt{t\, A\, \eta} \end{aligned}$$

Q.E.D.

Therefore, the exposure time necessary to reach a given SNR goes as

$$t \propto \frac{1}{A\eta}. \tag{3.4}$$

Thus, doubling the efficiency of an instrument will halve the exposure times, and doubling the diameter of a telescope will quarter the exposure times.

The signal-to-noise ratio gives a measurement of the significance of an observation based on the Poisson noise due to the various background contributions. However, the accuracy of any photometric observation is usually a combination of Poissonian errors and systematic errors.

3.1.1.1 *Systematic errors*

Systematic errors may be considerable, especially for long exposures. For example, flexure, and thermal instability can cause the instrument response to change. The sky background can change with zenith distance, distance from the moon, and at certain wavelengths is intrinsically variable. The consequence of systematic errors is that the signal-to-noise cannot be indefinitely increased through long exposures; there are practical limits on exposure time depending on the stability of the instrument and the background.

The systematic errors in an observation can be minimised by ensuring that the instrument is as stable as possible, and that the instrument throughput is as high as possible. A high throughput will make the contributions to the noise from the photon counts higher, and thereby lessen the significance of other sources of error, as well as increasing the signal-to-noise overall. The consideration of throughput applies to all aspects of the instrument from the telescope to the detector, including insertion losses into any waveguides, which will be discussed in detail in Chapter 6. Typical end-to-end throughputs of imagers are $>50\%$.

Instability in the instrument and background can be calibrated, e.g. by observing standard stars of known magnitude. Furthermore, variations in the response of the instrument from pixel to pixel across the FOV can be calibrated by means of a flat-field observation, i.e. by observing a source of uniform surface brightness — usually either the twilight sky or a diffusing screen illuminated with a lamp. However, it is desirable that the response across the FOV be as flat as possible to begin with to avoid systematic and statistical errors in this calibration, especially since many observations are carried out at the limit of sensitivity, since this is naturally where many untested regions of parameter space lie.

3.2 Imagers

3.2.1 *Scientific applications*

Imaging of celestial objects remains an indispensable part of astronomy. Imaging allows the study of the positions, sizes, distributions, morphologies and brightnesses of various objects, and enables searches for transient events. The scientific outcomes of such studies are fundamental to all of observational astronomy.

Figure 3.1 shows an image of a distant cluster of galaxies, as an example of the sort of information even a simple image can contain. This image is made in the K band at a wavelength of $\approx 2.2\,\mu$m. Information in this image can yield the presence of a cluster of galaxies from the overdensity of galaxies in the image, the distribution of these objects indicates the mass profile of the cluster, the brightness of the galaxies indicates their

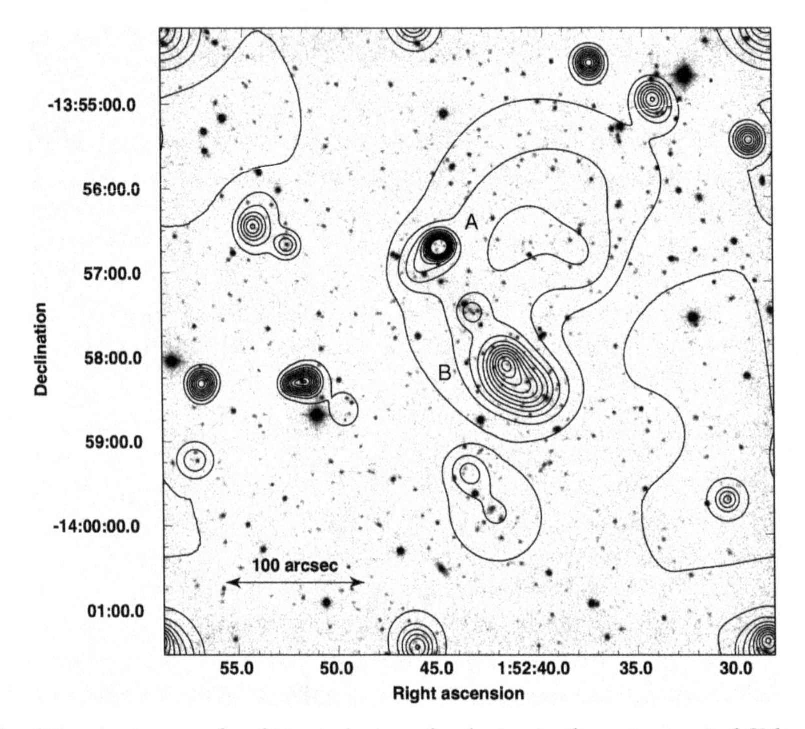

Fig. 3.1. An image of a distant cluster of galaxies in the astronomical K band (a wavelength $\approx 2.2\ \mu m$). Almost every object visible is a galaxy. The contours show the number density of galaxies in the image. In this case there are two distinct sub-clusters, labelled A and B, which are in the process of merging. Even such a simple image can provide information on the formation and build-up of structure in the Universe, and details of the evolution of galaxies. The image was made with the IRIS2 instrument on the Anglo-Australian Telescope. Reproduced with permission from Maughan *et al.* (2006). © The American Astronomical Society.

stellar mass, which can be compared to nearby systems to infer details of the evolution of galaxies.

3.2.1.1 *Photometry, passbands and filters*

Photometry, the measure of the brightness of astronomical sources, is one of the most common and important uses of imagers, since the brightness of an object is one of the most fundamental astronomical quantities. Strictly, imaging is not necessary for photometry, and historically photometry was often made using photomultiplier tubes which had no imaging capability, but which were more sensitive than photographic plates. Today, with the extremely high and uniform QE of astronomical imaging arrays photometry is more usually carried out in conjunction with imaging.

Detectors are sensitive over a wide wavelength range (Figure 2.9), but photometric observations are usually made in a particular passband, defined as a combination of an agreed upon set of filters and the transmission of the Earth's atmosphere, see Figure 3.2. Well defined passbands allow proper comparison between different observations. In addition to broad-band filters such as those shown in Figure 3.2, many specific science cases require narrow-band filters to isolate particular spectral features, e.g. specific emission lines. A series of many narrow band filters at adjacent wavelengths can be used to obtain a coarse spectrum of each object in the field, which can give rough classifications of objects and approximate redshifts.

In optical and near-infrared astronomy the brightness of an object is often expressed as an **apparent magnitude**, m_p, in a particular band, where

$$m_\mathrm{p} = -2.5 \log F_\mathrm{p} + Z_\mathrm{p}, \tag{3.5}$$

and F_p is the flux of the object within that band and Z_p is a constant which is used to calibrate the measured magnitude to a standard photometric

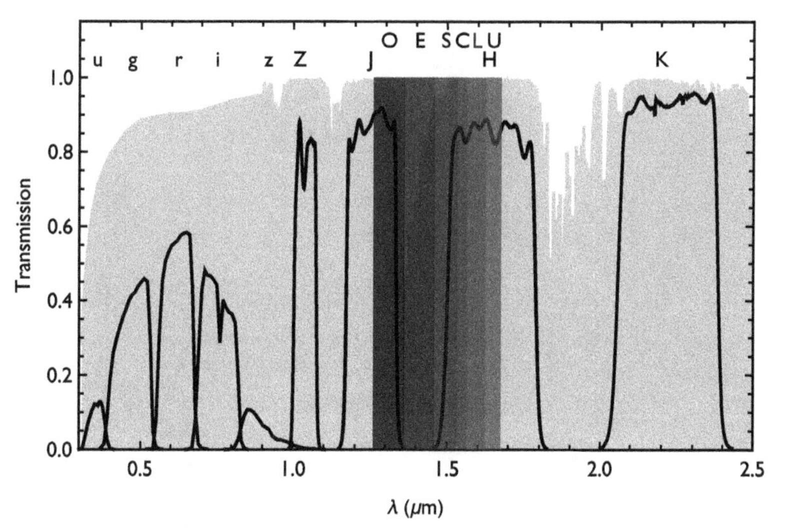

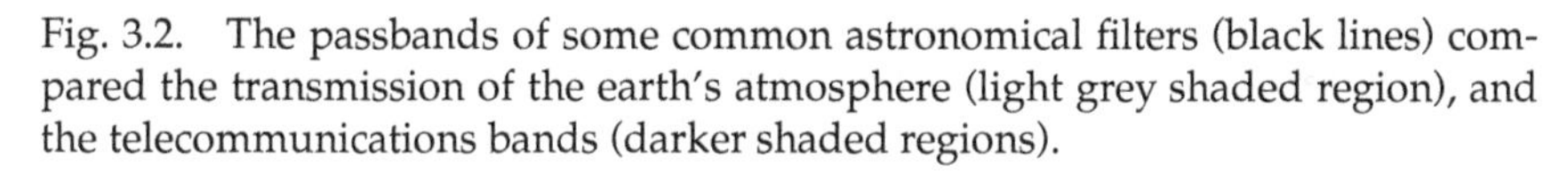
Fig. 3.2. The passbands of some common astronomical filters (black lines) compared the transmission of the earth's atmosphere (light grey shaded region), and the telecommunications bands (darker shaded regions).

system, and which must be determined from observations of standard stars of known flux taken during the same night and under similar conditions as the observations of the object in question.

If the flux is given in Jansky (Jy, i.e. 10^{-26} W m^{-2} Hz^{-1}), then the **AB magnitude** is given by

$$m_{\mathrm{AB}} = -2.5 \log F_{\mathrm{p}} + 8.9, \tag{3.6}$$

$$= -2.5 \log \left(\frac{F_{\mathrm{p}}}{3631} \right), \tag{3.7}$$

in any band, i.e. the flux of a 0 AB mag object is ≈3631 Jy. Another commonly used system is **Vega magnitudes** in which the magnitude of the star Vega (of class A0V) is defined to be zero in all passbands. We will discuss conversions between these two systems.

Alternatively, the brightness of a source could be expressed in physical units. For example, the intrinsic luminosity of a source expressed in W, or more commonly in optical astronomy in erg s^{-1}, with 1 erg = 10^{-7} J. The flux of a source is

$$F = \frac{L}{4\pi d_{\mathrm{L}}^2}, \tag{3.8}$$

where L is the luminosity and d_{L} is the luminosity distance to the source,[1] and is often given in units of erg s^{-1} cm^{-2}. Either of these may also be expressed as a surface brightness, e.g. flux per square arcsec.

It is often necessary to convert between the counts recorded on a detector (in ADU, or analogue to digital units) to a physical quantity such as flux, for example,

$$\mathrm{ADU} = F \times \frac{\lambda}{hc} \times A_{\mathrm{tel}} \times t_{\mathrm{exp}} \times g \times \eta, \tag{3.9}$$

where $\frac{hc}{\lambda}$ is the energy of a photon of wavelength λ, A_{tel} is the collecting area of the telescope, taking into account any central obstruction, t_{exp} is the exposure time, g is the gain of the detector in ADU/e$^-$, and η is the total throughput of the optical train from the top of the atmosphere to the detector.

[1] In a flat and static space the luminosity distance is simply the usual Euclidean distance; however, in a cosmological context the luminosity distance also depends on the curvature of spacetime, the redshift due to the expansion of the Universe, and time dilation.

Table 3.1. Photometric parameters of common photometric passbands, including Johnson, SDSS, Mauna Kea NIR, and IR filters.

Passband	Central λ (μm)	Bandpass (μm)	C_0 (10^{10} ph/s/m^2/μm) AB	Vega	Vega–AB (mag)
Johnson U	0.365	0.068	15.014	7.2524	0.79
Johnson B	0.440	0.098	12.454	13.531	−0.09
Johnson V	0.550	0.089	9.9635	9.7817	0.02
Johnson R	0.700	0.220	7.8285	6.4517	0.21
Johnson I	0.900	0.240	6.0888	4.0228	0.45
SDSS u	0.356	0.046	15.393	6.6577	0.91
SDSS g	0.483	0.099	11.346	12.213	−0.08
SDSS r	0.626	0.096	8.7539	7.5544	0.16
SDSS i	0.727	0.106	7.1446	5.0814	0.37
SDSS z	0.91	0.125	6.0219	3.6621	0.54
MK J	1.215	0.26	4.5102	1.9507	0.91
MK H	1.654	0.29	3.3131	0.9210	1.39
MK K	2.179	0.41	2.5149	0.4576	1.85
L	3.547	0.57	1.5450	0.117	2.80
M	4.769	0.45	1.1491	0.0506	2.29
N	10.472	5.19	0.5233	0.0051	5.03
Q	20.13	7.8	0.2722	0.0007	6.43

Note: The values of C_0 are for objects of 0 AB magnitudes. The final column of Table 3.1 gives the conversion between AB and Vega magnitudes for each band.

To convert between magnitudes and physical units it is necessary to know the constants C_p for each pass band. These are given in Table 3.1 for some common filters for both the AB and Vega magnitude systems, expressed as C_0 which is the flux of a 0th mag object in photons s^{-1} m^{-2} μm^{-1}. Using these values, it is possible to convert between a magnitude, m, and a physical flux using,

$$m = -2.5 \log \frac{F}{C_0}. \tag{3.10}$$

3.2.1.2 *Astrometry*

Astrometry is the measurement of the positions of celestial objects. This is a very ancient endeavour, with astrometric catalogues of the brightest stars dating back to at least the 2nd century BCE and the early star catalogues compiled and analysed by Hipparchus and later by Ptolemy.

Astrometry has had a continued and enormous impact on science throughout history, from Hipparchus' discovery of the Earth's precession, to celestial navigation, and measurement of the distances to the closest stars by trigonometric parallax. Astrometry remains fundamental to astrophysics today. For example repeated astrometric observations of stars at the Galactic centre have revealed the unambiguous presence of a supermassive black hole, and tested the predictions of general relativity (Gravity Collaboration *et al.* 2018, 2019); similar observations of the motions of stars can be used to search for the presence of black holes in globular clusters. Large scale astrometric programmes are fundamental to the understanding of the formation of our own Galaxy (Gaia Collaboration *et al.* 2016).

Within the context of imagers, astrometry will usually be relative, i.e. the separations of objects within the image will be measured, rather than their absolute position on a world coordinate system. For more details on absolute astrometry we refer the reader to Johnston and de Vegt (1999).

In order to perform relative astrometry within an image it is necessary to know precisely the plate-scale at the detector (see Section 2.2.4), and also any distortions. All aberrations, such as de-focus, coma, astigmatism etc. will affect the accuracy of astrometry, but distortions in particular must be modelled and corrected to perform astrometry. Illustrations of barrel and pincushion distortions are shown in Figure 3.3. Such distortions can be calibrated from sources of known absolute astrometry, or from a grid of pinholes placed at an intermediate image plane within the instrument.

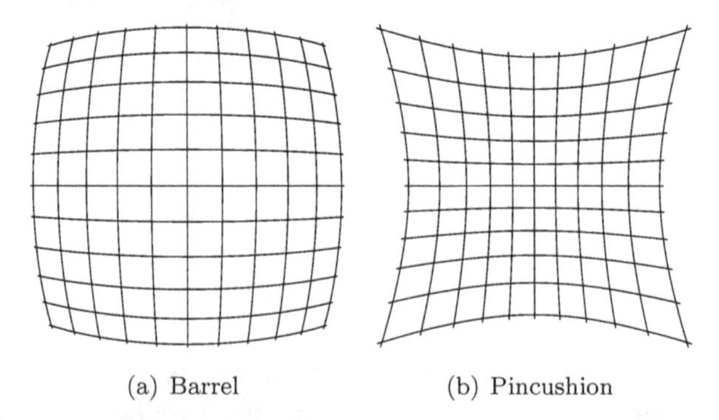

(a) Barrel (b) Pincushion

Fig. 3.3. Sketches of barrel and pincushion distortions, in which the plate-scale varies across the FOV; a rectangular grid of straight lines appears distorted at the image plane. Such distortions must be calibrated in order to perform astrometry within an image.

3.2.2 *Instrumentation and requirements*

At its most basic, an imager can consist of a detector at the focus of a telescope. However, as discussed in Section 2.2.4, it is usually required to reimage the telescope focus via fore-optics such as shown in Figure 2.6(b). Such re-imaging optics have several advantages. First, they allow the selection of the desired plate-scale, appropriate to the scientific requirements (see Section 2.2.4). Second, they introduce a collimated beam in which to place any elements such as filters, and the filters can usually be made smaller and therefore cheaper. Third, they produce a pupil image, at which an aperture stop can be placed to baffle any extraneous scattered light or thermal emission from the instrument.

Figure 3.4 shows the design of the IRIS2 imaging spectrograph (Gillingham and Jones 2000; Tinney *et al.* 2004) from the Anglo-Australian Telescope, which was used to make the image in Figure 3.1, compared to a sketch of a basic imager, cf. Figure 2.6(b). The collimator and camera lenses consist of a series of lenses of various material, chosen to minimise chromatic and spherical aberrations at the operating wavelengths of the instrument. IRIS2 can act as an imager or a spectrograph, and so has both a set of filters and grisms which can be inserted or removed from the collimated beam.

The design of any instrument consists of a series of trade-offs between competing features, including the scientific benefit, the available technology, and the cost. Typical design choices are discussed as follows.

3.2.2.1 *Detectors*

Detectors have been discussed in detail Section 2.6. The selection of an appropriate detector is of primary consideration when designing an imager. In particular, the QE as a function of wavelength should be appropriate for the application. Typically in astronomy, this means as broad a range as possible over the entire visible wavelength range (350–1,000 nm for Si CCDs) or near-infrared wavelength range (1,000–2,500 nm for HgCdTe or 1,000–5,000 nm for InSb detectors), although in particular applications certain wavelengths may carry more importance. Note that these wavelength ranges are very broad compared to typical photonic applications. Ideally, one would like to be able to select the pixel size appropriate for the telescope or instrument optical design. In practice however, detectors are available with fixed pixel sizes, typically

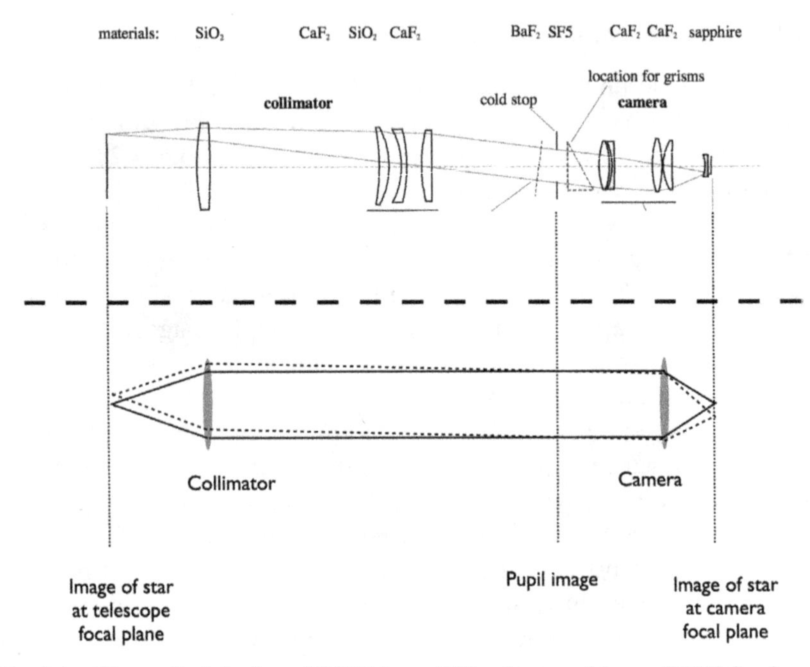

Fig. 3.4. The optical design of IRIS2 from Gillingham and Jones (2000) (top) compared with a sketch of a basic collimator-camera (cf. Figure 2.6).

10–18 μm, and the image must be magnified to the appropriate plate-scale as described in Sections 2.2.4 and 2.3.

3.2.2.2 *Field-of-view and plate-scale*

The FOV of an imager should be large enough to accommodate typical objects, if their sizes and morphologies are to be measured (e.g. large nearby galaxies). Likewise, if measuring the distribution of objects, then the FOV should be large enough to observe a sufficient number of objects in a single exposure such that observations may be made in a reasonable amount of time.

The necessity to choose an appropriate pixel size was discussed in Section 2.2.4. Typically there should be two pixels per resolution element to satisfy Nyquist sampling. However, if high spatial resolution is a priority then oversampling the point spread function (PSF) will be beneficial, at the cost of decreased sensitivity. If sensitivity is a priority then undersampling the PSF (PSF, i.e. the size of a point source at the detector) may be beneficial, at the cost of lost resolution.

For a specific detector, changing the plate-scale will necessarily change the FOV. A larger plate-scale (more arcsec per mm) will result in each pixel covering a larger area of the sky and a larger FOV, along with possible undersampling.

Of course one can always increase the FOV by using multiple or larger format detectors, and ultimately the FOV of most imagers is limited by the cost of detectors and the number of pixels available.

Very wide field imagers require more complicated optics to reduce the spherical aberrations from highly off-axis objects. For example, Figure 3.5 shows the design of the Hyper Suprime-Cam, a 1.5° camera being built for the prime focus of the 8.2 m Subaru telescope.

3.3 Spectrographs

3.3.1 *Scientific applications*

Spectroscopy is the cornerstone of astrophysics. A spectrum of an object is capable of yielding information about its chemical composition, its radial velocity, its rotation, its temperature, its pressure, and its density. Some example spectra are shown in Figure 3.6.

Spectrometry can be combined with other measurements (Section 3.1). For example, spectrophotometry, combines spectrometry with photometry to yield the absolute brightness of the object per unit wavelength in physical units. This requires comparison with a known calibration source and accurate knowledge of the losses at the entrance slit to the spectrograph, since typically not all the light from the object will be captured. Spectroscopy can be combined with imaging as will be discussed in Section 3.3.4, and likewise with interferometry or polarimetry.

3.3.2 *Instrumentation and requirements*

The applications of spectroscopy are so diverse that there are many different requirements and designs of spectrograph. Figure 3.7 shows a sketch of the components of a basic spectrograph. Light from the telescope is focussed onto a slit, such that only light from the object of interest is passed onto the spectrograph. In much of our discussion the 'slit' will actually be the outputs from a linear array of optical fibres. Note again the basic collimator-camera design from Figure 2.6: the diverging light is collimated

(a) Hyper Suprime-Cam Layout

(b) Hyper Suprime-Cam Detectors

Fig. 3.5. The wide field imager Hyper Suprime-Cam, a 1.5° imager for the prime focus of the 8.2 m Subaru Telescope. (a) The layout of the instrument including the wide field corrector necessary to correct the spherical aberrations which result from very large off-axis angles. (b) The detector array at the focal plane, comprised of 116 CCDs with 870 million pixels in total. Reproduced with permission from HSC Project.

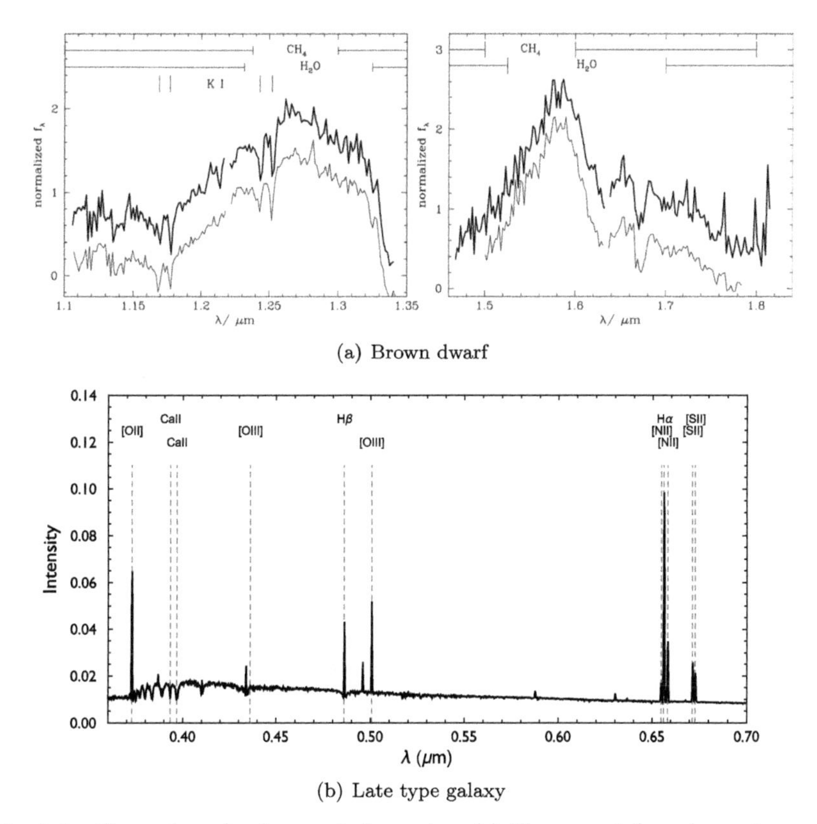

Fig. 3.6. Examples of astronomical spectra. (a) The near-infrared spectrum of a brown dwarf star. The relative strength of the molecular absorption features allows the star to be classified as a type T5 dwarf. The thin grey spectrum shows a prototype T5 dwarf for comparison. Reproduced with permission from Ellis *et al.* (2005). © The American Astronomical Society. (b) A template spectrum of a late type galaxy from the Sloan Digital Sky Survey (http://classic.sdss.org/dr5/algorithms/spectemplates/). The wavelengths of the emission lines allows its red-shift to be measured. The strength of the hydrogen and oxygen lines allow the star-formation rate in the galaxy to be estimated. The relative strength of the hydrogen, nitrogen and sulphur line allow the ionisation state of the gas to be measured.

with a lens or a mirror, and then dispersed, e.g. with a diffraction grating or a prism. Finally, the dispersed light is focussed onto a detector to record the spectrum. Many spectra from multiple slits or fibres can be recorded side-by-side on the detector.

As an example of a real instrument, Figure 3.8 shows the optical design of the HERMES spectrograph (Sheinis *et al.* 2015) at the Anglo-Australian

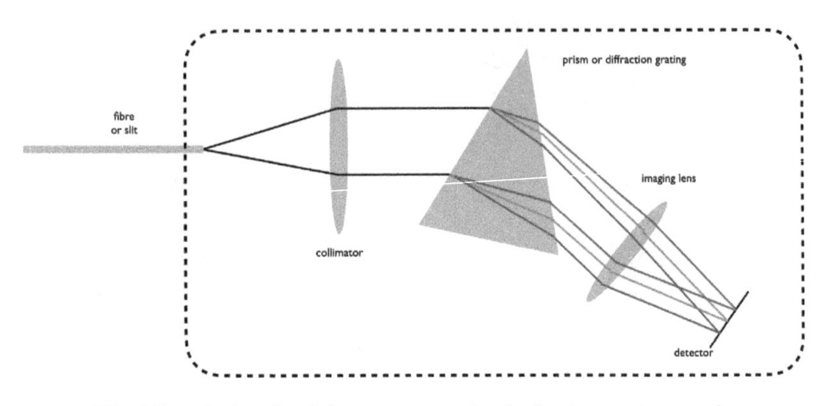

Fig. 3.7. A sketch of the components of a basic spectrograph.

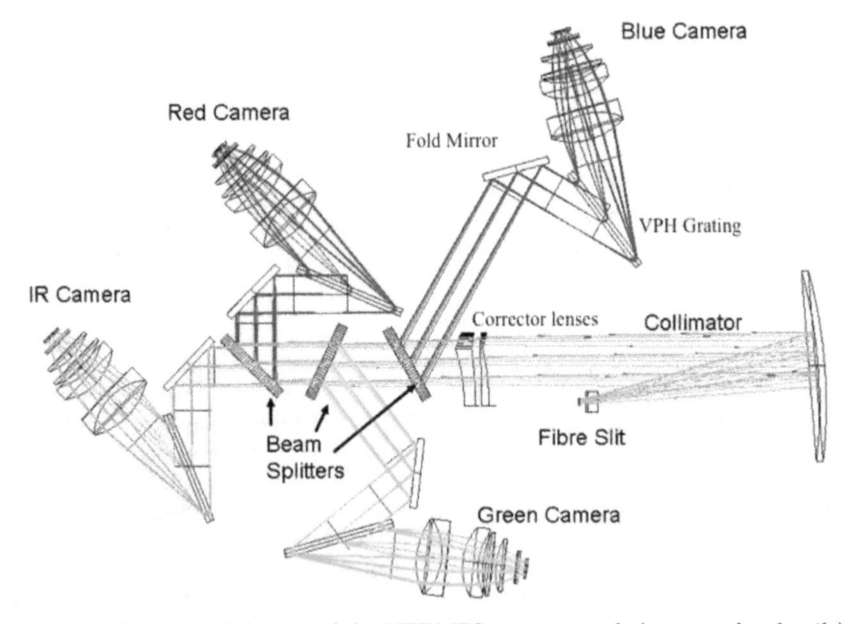

Fig. 3.8. The optical design of the HERMES spectrograph (see text for details). Reproduced with permission from Sheinis *et al.* (2015). © Creative Commons licence (CC BY 3.0).

Telescope. The light enters the spectrograph from 400 fibres aligned along a pseudo-slit. The light from these fibres is reflected off a collimating mirror, shown at the right hand side of the diagram. After some lenses to correct spherical aberrations, a series of three dichroic beam splitters splits light of different wavelengths into four separate arms, each of which contains a similar camera. In each camera, the light is first reflected off a fold mirror, which reduces the physical size of the spectrograph. Thereafter, the light is diffracted by a volume phase holographic grating, whereupon it is focussed by the camera, consisting of a series of lenses, onto a CCD.

Again, we limit our discussion to the most basic design choices, which are relevant to inform the discussion of the astrophotonic instruments, and to bring attention to the differences between astronomical requirements and photonic requirements.

3.3.2.1 *Resolving power and wavelength range*

Due to the limited number of pixels available at the detector, a major design choice is the trade-off between wavelength range and **resolving power**, R, where

$$R = \frac{\lambda}{\Delta\lambda}, \tag{3.11}$$

and $\Delta\lambda$ is a suitable measure of the resolution, e.g. the full width at half-maximum (FWHM) of a spectral line. For high resolving power the wavelength range will be smaller, and vice versa. The resolution, $\Delta\lambda$, should again be Nyquist sampled in order to ensure the proper reconstruction of the spectrum.

Example 3.3: Consider a spectrograph with a minimum resolving power of $R = 1{,}000$, and a minimum wavelength of 400 nm. If the CCD has 2,048 pixels in the spectral direction and the PSF is sampled by 2 pixels, what is the maximum wavelength, and what is the resolving power at this wavelength? How does this change if the resolving power is $R = 10{,}000$?
The resolution element is $\Delta\lambda = 400/1{,}000 = 0.4$ nm. If this is sampled by 2 pixels, each pixel then covers 0.2 nm. The maximum wavelength is then $400 + 0.2 \times 2{,}048 \approx 810$ nm. At this wavelength the resolving power is $R = 810/0.4 = 2{,}025$.

At a minimum resolving power of $R = 10{,}000$, the maximum wavelength becomes 441 nm, and the maximum resolving power is $R \approx 11{,}000$.

Note well that if the spectrograph slit is narrower than the FWHM of the image projected onto it, then $\Delta\lambda$ is determined by the width and shape

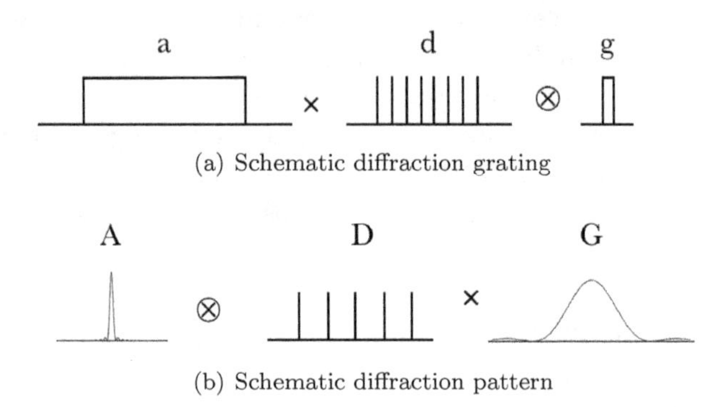

(a) Schematic diffraction grating

(b) Schematic diffraction pattern

Fig. 3.9. (a) A schematic representation of a diffraction grating and (b) the resulting Fraunhofer diffraction pattern, following Hutley (1982).

of the slit, and not by the dispersion. In the case of optical fibres forming a slit, then $\Delta\lambda$ is always determined by the diameter of the fibres. For example, consider Figure 3.9 representing a rectangular slit and a rectangular groove profile, following Hutley (1982). We may consider an ideal diffraction grating to be composed of a Dirac comb, d, convolved with the groove function of the grating, g. This is illuminated by the aperture function, a, which is the pupil of the spectrograph imaged onto the grating. Thus, the diffraction grating may be described as

$$\text{diffraction grating} = a \times (d \otimes g), \tag{3.12}$$

as illustrated in Figure 3.9(a). The Fraunhofer diffraction pattern of this grating is thus given by the Fourier transform of Equation 3.12, i.e.

$$\text{diffraction pattern} = (A \otimes D) \times G, \tag{3.13}$$

where A is the Fourier transform of a, etc., as in Figure 3.9(b). This simple picture makes it immediately apparent that the PSF (given by A) of any order (given by D) is due to the shape of the aperture function, i.e. it is determined by the width and shape of the slit. The shape of the grooves (or more generally the diffraction grating function) determines the relative brightness of each order, e.g. a blazed groove function would put more light into a higher order.

The above description is merely a pictorial representation of the familiar **grating equation**, where the irradiance is given by,

$$I = \frac{\sin^2\beta}{\beta^2}\frac{\sin^2 N\gamma}{\sin^2\gamma}, \tag{3.14}$$

where

$$\beta = \frac{kb}{2} \sin\theta; \quad \gamma = \frac{ka}{2} \sin\theta, \tag{3.15}$$

and $k = 2\pi/\lambda$, a is the slit separation, b is the slit width, and θ is the angle from the normal of light leaving the grating. Thus, the $\text{sinc}^2\beta$ term is the irradiance pattern due to a single slit (or groove) of the grating, i.e. it is equivalent to G above, and the second term, involving γ, is the interference due to the illumination of N lines on the grating, i.e. it is equivalent to A above.

A specific resolving power may be necessary to allow certain measurements, e.g. to resolve a specific line width to measure velocity dispersions, or to separate two close lines. On the other hand, for certain science goals, such as measuring the redshift of galaxies it may be more important to have a wider wavelength range, at the expense of unresolved features.

A spectrograph with a higher resolving power will have a relatively lower signal-to-noise for a given object, since a given amount of light is being spread over more pixels, and thus, there are fewer photons per pixel for a given exposure time.

The typical wavelength range of astronomical instruments is very high compared to many other photonic applications. Figure 3.10 shows the wavelength range, w, as a function of the resolving power and central wavelength, assuming a sampling of 2 pixels per FWHM of the line profile. Wavelength ranges of hundreds of nm are common.

In **wavelength division multiplexing (WDM)** applications within telecommunications a waveguide may transport several signals, each of a different wavelength. These are then divided at the output into their separate signals using a wavelength selective device. In many cases, such as arrayed waveguide gratings (Section 9.2), this technique is very similar to astronomical spectroscopy. In astronomical spectrographs however, the wavelength range across the detector is almost always continuous, such that the entire spectrum over the wavelength range of the detector can be reconstructed, rather than a measurement of the spectrum at discrete wavelengths. On the other hand, not all the information contained in a spectrum is necessarily useful. For example, if one wishes to measure the star-formation rate in a galaxy, only those pixels containing the relevant emission lines (and a few pixels around these for continuum subtraction) are required.

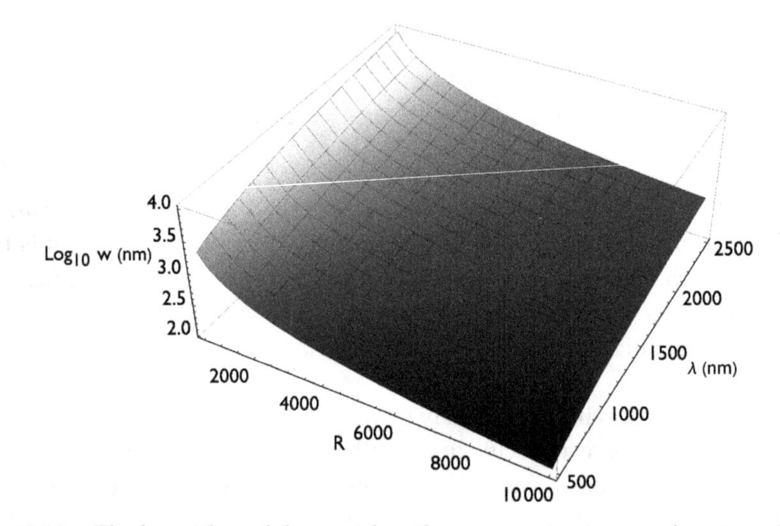

Fig. 3.10. The logarithm of the wavelength range, w, in nm, as a function of the resolving power and central wavelength, assuming a sampling of 2 pixels per FWHM of the line profile.

3.3.3 *Multi-object spectroscopy (MOS)*

It is possible to record the spectrum of many objects simultaneously with the same spectrograph. For example, if the entrance slit of a spectrograph is replaced with a mask containing multiple short slits, each located at the position of an object of interest, then a spectrum of each object can be recorded, as sketched in Figure 3.11(a). If the slit is longer than the object, then the sky spectrum, which is necessary for background subtraction, can be simultaneously recorded. Depending on the location of the slits, not all spectra may cover the full wavelength range, and thus the FOV is limited by the size of the detector. Furthermore, care must be taken that spectra do not overlap.

Conversely, for a fibre-fed spectrograph the outputs of each fibre can be aligned such that all spectra record the full wavelength range and have an identical format, see Figure 3.11(b). Moreover, the FOV can be much larger than for a slit-mask, as fibres can be located anywhere on the focal plane. Since it is not possible to simultaneously record sky and object spectra in a single fibre, some fibres must be dedicated to sky observations, or else separate sky exposures must be taken.

In fact, multi-object spectroscopy with an array of fibres was the first application of astrophotonics, and indeed remains the most common

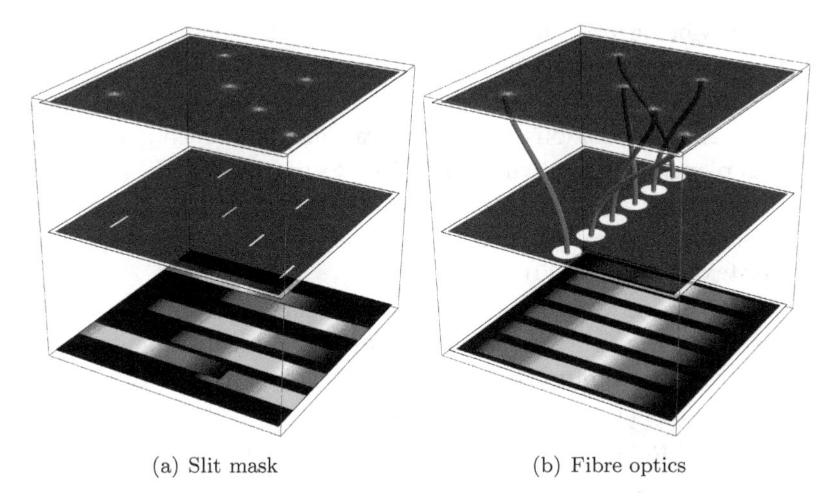

Fig. 3.11. (a) An outline of multi-object spectroscopy with a slit-mask and (b) with fibre optics, showing the image at the focal plane (top plane), the slit mask (middle plane) and the spectra at the detector (bottom plane), and optical fibres in blue. In the mask each slit is located at the position of the focal plane corresponding to an object of interest, and so the spectra are also distributed on the detector in the same pattern. This means that the full wavelength range cannot always be recorded without limiting the FOV, and care must be taken that the spectra from different objects do not overlap. In the case of the fibres, the output of each fibre can be aligned, such that all spectra are fully recorded and have the same format. The FOV can be made much larger than the slit mask.

application of waveguides in astronomy. Despite the relatively basic (from a photonics point-of-view) use of fibres as light pipes in multi-object spectroscopy, fibre-fed MOS transformed extragalactic astronomy, affording surveys of millions of galaxies (e.g. the Sloan Digital Sky Survey, SDSS, and the 2dF Galaxy Redshift Survey), and constraining the large-scale structure of the Universe and the cosmological parameters governing its evolution.

Both fibre fed spectrographs and slit masks can be reconfigurable. The position of fibres on the focal plane can be reconfigured using robotic fibre positioners, or by hand using plug-plates (i.e. a metal plate with pre-drilled holes at the position of the objects, into which the fibres are plugged). Typically this is done using pick-and-place robots to reposition one fibre at a time (e.g. Dalton *et al.* 2018; Lewis *et al.* 2002). Since, this is time consuming for large numbers of fibres, many such instruments have duplicate focal planes, which allows one to be reconfigured whilst the other is being used for observations. Simultaneously, reconfiguring is possible with individual fibre positioners such as 'echidna spines' (e.g. Gillingham *et al.* 2000), 'theta-phi' positioners (e.g. Xing *et al.* 1998), or 'starbugs' — semi-autonomous robots that can move independently around the focal plane (e.g. Goodwin *et al.* 2010). Reconfigurable slit masks have been made using an array of small slitlets, each of which can be opened or closed independently, to achieve the desired configuration (e.g. Henein *et al.* 2004; McLean *et al.* 2010).

The maximum number of objects which can be observed simultaneously is ultimately limited by the number of available pixels. In the case of a slit mask the maximum is set by the requirement that spectra must not overlap, and must fall onto the detector, as discussed above. In the case of fibre fed spectrographs the maximum is set by the number of fibres which can fit along the entrance slit of the spectrograph whilst leaving a sufficient gap between adjacent fibres such that the spectra do not overlap on the detector.

3.3.4 *Integral field spectroscopy (IFS)*

An **integral field spectrograph** is a particular type of multi-object spectrograph in which the spectra are sampled from a contiguous region at the focal plane. This allows spectra from different parts of the same object to be taken simultaneously. This is very useful in many areas of astrophysics. For example, IFS of galaxies allows the velocity of a galaxy to be measured at different points, which can measure the rotation of the galaxy and the velocity dispersion, which are essential for understanding the internal dynamics of galaxies. Similarly, measurements of specific emission lines can measure the star-formation rate, the metallicity, and hence stellar ages, the ionisation state, etc. as a function of position throughout the galaxy, allowing details of galaxy evolution to be studied. For a thorough and up-to-date treatise on integral field spectroscopy and spectrographs, including references, see the excellent book by Bacon and Monnet (2017). Figure 3.12 shows an example observation of a Seyfert galaxy with the KOALA integral field spectrograph at the AAT (Zhelem *et al.* 2014).

The component of the instrument which divides the focal plane and feeds each segment to the spectrograph is known as an **integral field unit (IFU)**. There are several ways to make an IFU. Clearly a simple slit mask, as in Figure 3.11(a), will not work as the spectra will overlap one another. Instead the slit-mask can be replaced with a lenslet array, as in Figure 3.13(a). Each lenslet will project a small pupil image onto the entrance of the spectrograph, and these pupil images can be adequately separated from each other. By tilting the lenslet array with respect to the direction of the dispersion, the spectra can thus be made to avoid overlapping. Nevertheless, spectra from the edges of the lenslet array may not be fully recorded on the detector, and there are significant areas of unused pixels on the detector.

The focal plane can also be divided with an array of fibres. In this case, the detector can be used more optimally since the output from each fibre can be aligned to ensure full wavelength coverage from each element, see Figure 3.13(b), as was the case for MOS, cf. Figure 3.11.

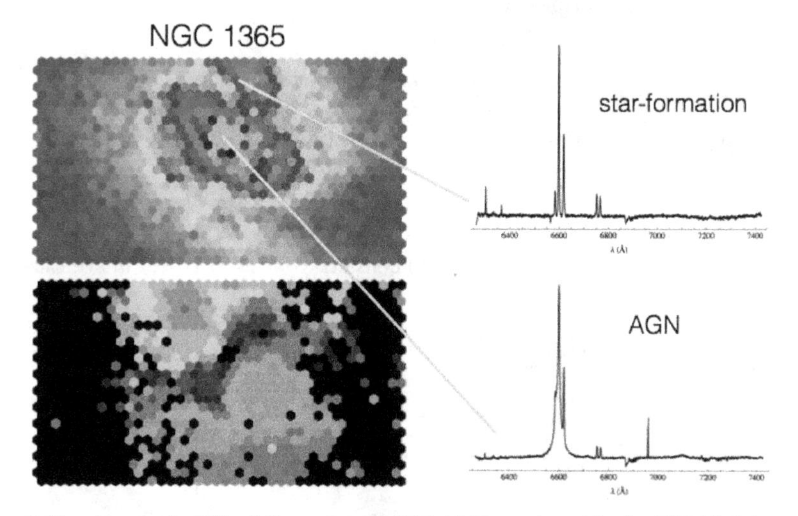

Fig. 3.12. Example IFS of the galaxy NGC 1365 made with the KOALA instrument on the AAT. The top left panel shows an image of the Hα emission from the galaxy, and the bottom image shows the velocity of the gas emitting the Hα radiation, with red points moving away from us, and blue towards us; the galaxy is rotating. The spectra on the right hand side show the difference in emission lines from different parts of the galaxy. The outer region shows emission lines and widths associated with star-formation regions. The central region shows the broad emission lines associated with gas accreting onto a central supermassive black hole, characteristic of an active galactic nucleus (AGN).

Example 3.4: Consider a hexagonally packed fibre array with fibres with a core diameter of 100 μm and a cladding of 250 μm. Estimate the filling-factor.

Figure 3.14 shows the layout of part of the fibre array, which can be seen to be composed of a number of similar equilateral triangles. The area of one triangle is

$$A_{\text{tri}} = \frac{\sqrt{3}}{4} d_{\text{clad}}^2, \tag{3.16}$$

where d_{clad} is the diameter of the cladding. Meanwhile, the combined area of the segments of the three fibre cores which fall within this triangle is

$$A_{\text{core seg}} = \frac{\pi}{8} d_{\text{core}}^2, \tag{3.17}$$

where d_{core} is the core diameter. Therefore, the approximate filling factor, neglecting the edges of the array, is

$$f = \frac{\pi}{2\sqrt{3}} \left(\frac{d_{\text{core}}}{d_{\text{clad}}} \right)^2, \tag{3.18}$$

which for the present example gives $f \approx 14.5\%$.

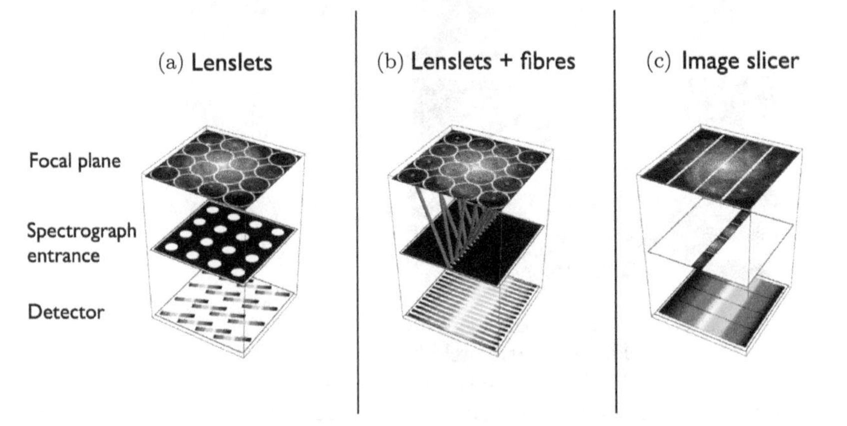

Fig. 3.13. Schematic working of three main types of integral field spectrograph, using (a) lenslet array; (b) lenslets and fibres; and (c) an image slicer, to divide the focal plane. The top plane shows how the image is dissected with lenslets, lenslets and fibres, or slicer-mirrors. The middle plane shows the entrance at the spectrograph, and the bottom plane depicts the output of the spectrograph onto the detector.

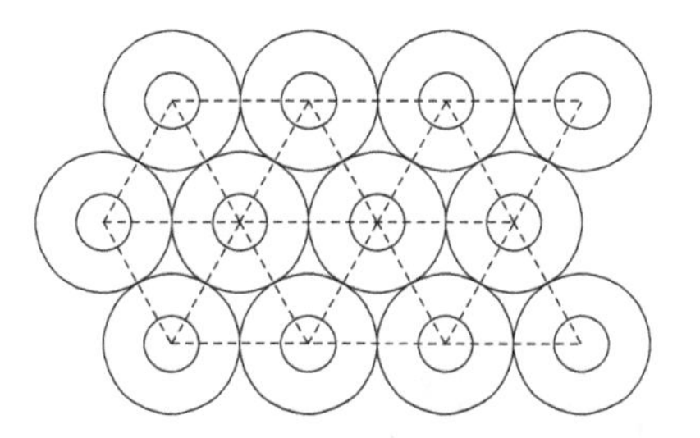

Fig. 3.14. Sketch of part of a hexagonally packed fibre array, illustrating the filling factor as in Example 3.4.

The filling factor can be made >95% by using a combination of microlens arrays and fibres. In this case, the microlens array has a contiguous coverage of the focal plane, and each microlens feeds one fibre. Lenslet arrays can be made of circular lenslets, in which case the maximum filling factor will be given by Equation 3.18 with $d_{\rm core} = d_{\rm clad}$, yielding $f \approx 91\%$. Higher filling factors can be achieved with square microlens arrays, and close to 100% filling can be achieved with hexagonal microlenses. A further advantage of using microlens arrays to feed the fibres is that it allows the fibres to be fed at the optimal focal ratio, and by projecting a pupil image onto the front-face of the fibre it is possible to ensure that all light is collected by the fibre — this will be discussed in detail in Section 6.2.

An alternative solution to making an IFU is to use image slicers, as depicted in Figure 3.13(c). In this case, the image plane is divided by a series of angled mirrors as sketched in Figure 3.15. A further two sets of mirrors aligns each of the sliced images onto the slit of the spectrograph. Image slicers are very efficient, and make highly optimal use of the detector, since there is no dead space between slices, as there must be between fibres.

A recent advance has been the development of deployable IFUs. These combine the advantages of MOS and IFUs, allowing IFS of multiple objects simultaneously. These can be composed of either image slicers or fibre arrays. In the first case, the slicers are fed using mirrors on movable pick-off arms. In the second case small bundles of fibres could be deployed using similar pick-and-place robots as in MOS (Cecil *et al.* 2010).

An interesting development in deployable IFUs is the **Hexabundle**. These are bundles of fibres which have been lightly fused together over a short distance. Because the fusing is light, the fibres retain their shape;

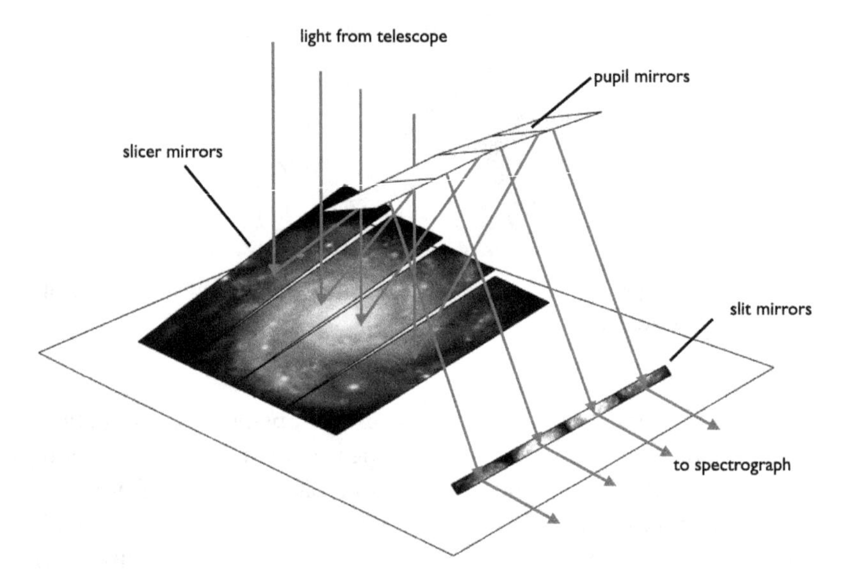

Fig. 3.15. A sketch of the principle of an image slicer IFU.

strongly fusing them would distort the cores and introduce unwanted focal ratio degradation (Section 4.2.5). The cladding in a hexabundle is very thin at the fused region to increase the filling-factor. Ordinarily this would lead to cross-talk in a fibre bundle, wherein the light from one fibre can leak into an adjacent fibre. However, provided the fused region is kept short the cross-talk has been shown to be negligible, even for thin cladding (Bryant *et al.* 2014).

3.4 Polarimeters

Astronomical polarimeters measure the degree of polarisation from a celestial source. Polarisation in astronomical sources typically occurs via scattering of the light, interaction with magnetic fields, or asymmetries of the sources, and its measurement therefore provides information on all these phenomena. It is measured either via imaging, or by spectroscopy, yielding the polarisation as a function of wavelength.

Consider an incoming elliptically polarised beam of light. The electric field may be decomposed into two orthogonal electric vectors given by

$$E_{\mathrm{x}}(t) = E_1 \cos(2\pi \nu t),$$

$$E_{\mathrm{y}}(t) = E_2 \cos(2\pi \nu t + \delta),$$

where E_1 and E_2 are the electric field amplitudes of each component, and δ is the phase difference between them, and ν is the frequency. The degree of polarisation is usually measured by the **Stokes' parameters** (see Kitchin 2009 for more details),

$$Q = E_1^2 - E_2^2, \tag{3.19}$$

$$U = 2E_1E_2 \cos\delta, \tag{3.20}$$

$$V = 2E_1E_2 \sin\delta, \tag{3.21}$$

$$I_\mathrm{p} = \sqrt{Q^2 + U^2 + V^2}, \tag{3.22}$$

$$I = I_\mathrm{u} + I_\mathrm{p}, \tag{3.23}$$

where I_u is the intensity of the unpolarised component of the light, and I_p is the intensity of the polarised component.

To see how these are measured we will use the example of the HIPPI-2 imaging polarimeter (Bailey *et al.* 2020), though many different schemes are possible. Figure 3.16 shows a schematic diagram of the instrument. Light incident on the instrument passes through a ferro-electric half-wave plate. This changes the handedness of any elliptical polarisation present, and is oscillated at 500 Hz. The Wollaston prism splits the light into its orthogonal polarisation components, the intensity of which are recorded in two separate photomultiplier detectors. The entire instrument is rotatable to different position angles.

The fractional polarisation can be measured from the signal in each detector D_1 and D_2, by

$$p = \frac{D_1 - D_2}{D_1 + D_2}. \tag{3.24}$$

The full normalised Stokes' parameters can be obtained from measurements at different position angles, and are given by

$$I = I(\theta) + I(\theta + 90), \tag{3.25}$$

$$\frac{Q}{I} = \frac{I(0) - I(90)}{I(0) + I(90)}, \tag{3.26}$$

$$\frac{U}{I} = \frac{I(45) - I(135)}{I(45) + I(135)}. \tag{3.27}$$

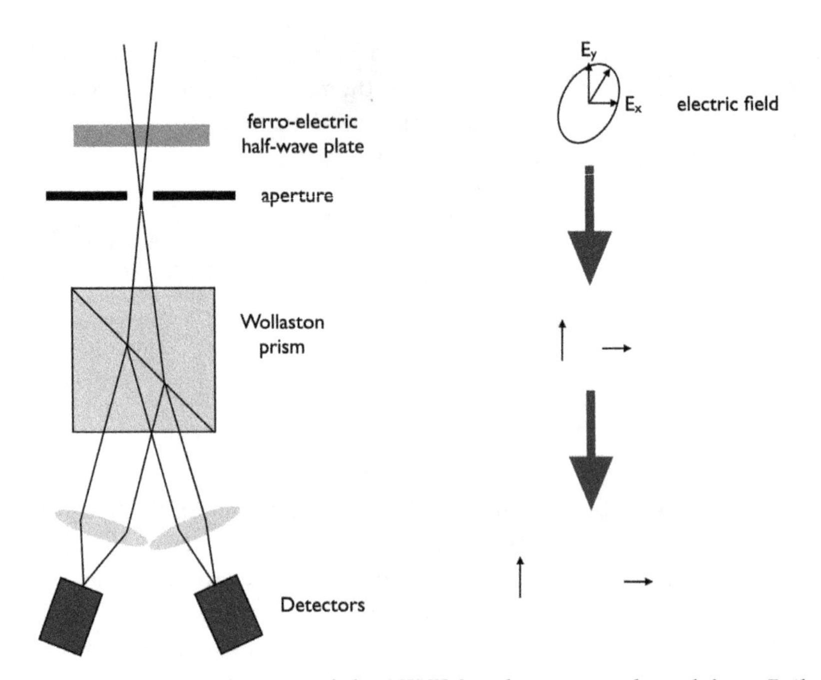

Fig. 3.16. Schematic diagram of the HIPPI-2 polarimeter, adapted from Bailey *et al.* (2020) and Kitchin (2009).

From which the full state of linearly polarised light at an angle ϕ to the x-axis can be found from

$$\frac{Q}{I} = \cos 2\phi,$$

$$\frac{U}{I} = \sin 2\phi.$$

3.5 Interferometers

Interferometers combine the beams from two or more separate apertures in order to produce an interference pattern. This has the advantage that the maximum angular resolution is determined by the separation of the apertures and not by the size of the apertures themselves (see Equation 2.1). This allows much higher angular resolution than in conventional imaging. For example, Figure 3.17 shows an image of a Wolf–Rayet star obtained using aperture masking interferometry (see Section 3.5.6) at the Keck Telescope. The spiral feature is a result of a very high velocity stellar wind

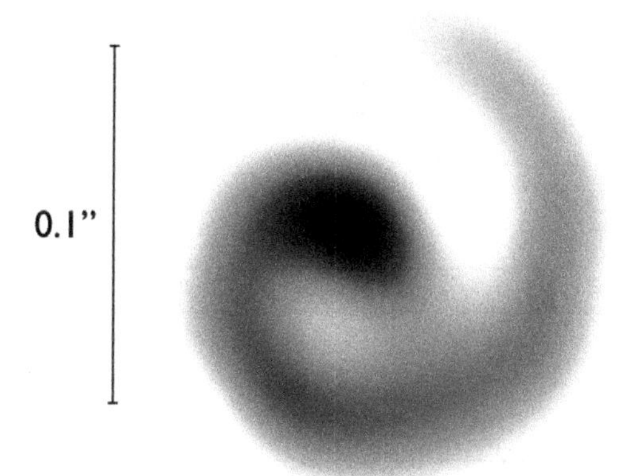

Fig. 3.17. An image of the Wolf–Rayet binary system, WR 104, obtained by aperture masking interferometry at the Keck Telescope. The spiral feature is dust formed in a shock front from the collision of the Wolf–Rayet stellar wind and the binary companion's stellar wind, which is driven outwards via radiation pressure. Reproduced with permission from Peter Tuthill, The University of Sydney and W.M. Keck Observatory.

arising from the extremely luminous Wolf–Rayet star. This wind interacts with the wind from a binary companion, and forms a shock front, in which dust is formed, which is driven outwards by the radiation pressure, resulting in the spiral surrounding the star.

Interferometry is a large field, with many different types of instruments and techniques. Here we concentrate only on those to which photonics have been applied, *viz.* beam transport and beam combination in optical interferometry and aperture masking interferometry (see Chapter 7). We give a brief overview of the basics of optical interferometry; for more information, see, e.g. Labeyrie *et al.* (2014) or one of many excellent radio astronomy books, e.g. Marr *et al.* (2015).

Interferometry requires the combination of beams from separate apertures in order to produce the interference. Consider Figure 3.18 showing a schematic diagram of an optical stellar interferometer. Light from a single point source is collected by two telescopes separated by a distance b.

The light arriving at the two telescopes is out of phase by

$$\phi = \frac{2\pi}{\lambda} b \sin\theta, \tag{3.28}$$

where θ is the zenith angle of the source, and we have made the simplifying assumption that the source is in the same plane as the telescopes and their zenith. The amplitudes of the light received at the two inputs to the

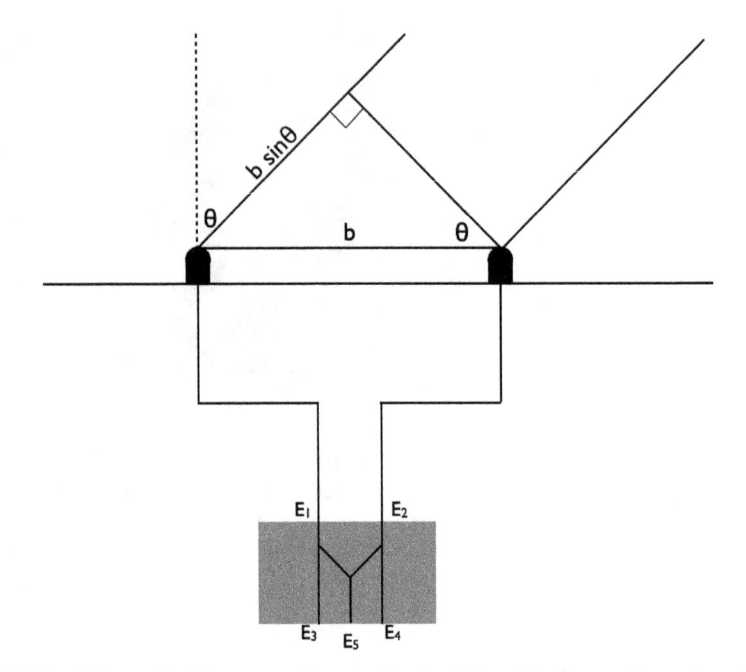

Fig. 3.18. Schematic diagram of an optical stellar interferometer observing a single point source.

beam combiner, E_1 and E_2, as a function of time are given by

$$E_1 = E_0 \cos 2\pi\nu(t + \tau), \tag{3.29}$$

$$E_2 = E_0 \cos 2\pi\nu t, \tag{3.30}$$

$$\tau = \frac{b \sin\theta}{c}. \tag{3.31}$$

Now consider the case in which the two signals are added together to produce the interference. In this case, the time averaged intensity at the output is

$$\begin{aligned} |E_5|^2 &= 2fE_0^2(1 + \cos 2\pi\nu\tau) \\ &= 2fE_0^2\left(1 + \cos\left(\frac{2\pi}{\lambda} b \sin\theta\right)\right), \end{aligned} \tag{3.32}$$

$$|E_3|^2 = (1 - f)E_0^2, \tag{3.33}$$

$$|E_4|^2 = (1 - f)E_0^2, \tag{3.34}$$

where $1 - f$ is the fractional intensity split from each beam to provide the photometric calibrations, E_4 and E_5.

Example 3.5: Derive Equations 3.32–3.34.

First, let us re-write Equations 3.29 and 3.30 in complex amplitude form

$$E_1 = E_0 e^{i2\pi\nu(t+\tau)}$$

$$E_2 = E_0 e^{i2\pi\nu t}.$$

Now, a fraction $\sqrt{f}$ of each *amplitude* is split off from each of E_1 and E_2 and added together to give E_5, i.e.

$$E_5 = \sqrt{f} E_0 \left(e^{i2\pi\nu t} + e^{i2\pi\nu(t+\tau)} \right).$$

Therefore, the intensity at E_5 is

$$\begin{aligned}
|E_5|^2 &= E_5 E_5^*, \\
&= f E_0^2 \left(e^{i2\pi\nu t} + e^{i2\pi\nu(t+\tau)} \right) \left(e^{-i2\pi\nu t} + e^{-i2\pi\nu(t+\tau)} \right), \\
&= f E_0^2 \left(2 + e^{i2\pi\nu\tau} + e^{-i2\pi\nu\tau} \right), \\
&= 2 f E_0^2 (1 + \cos 2\pi\nu\tau), \\
&= 2 f E_0^2 \left(1 + \cos\left(\frac{2\pi}{\lambda} b \sin\theta \right) \right). \text{ Q.E.D.}
\end{aligned}$$

Therefore, the intensity $|E_5|^2$ varies sinusoidally as the zenith angle, or the baseline separation b, changes.

3.5.1 *Aperture separation and size*

Now consider an interferometer looking at a closely separated pair of point sources as in Figure 3.19. Here we have introduced two delay lines. The first introduces a path length difference $\Delta L = -b \sin\theta$, such that the variation of the fringes due to the constantly changing zenith angle of the source is cancelled. The second path length difference δl is rapidly varied to scan through, or modulate, the fringes as a function of time. Now, the existence and frequency of the fringes is entirely due to the modulation δl,

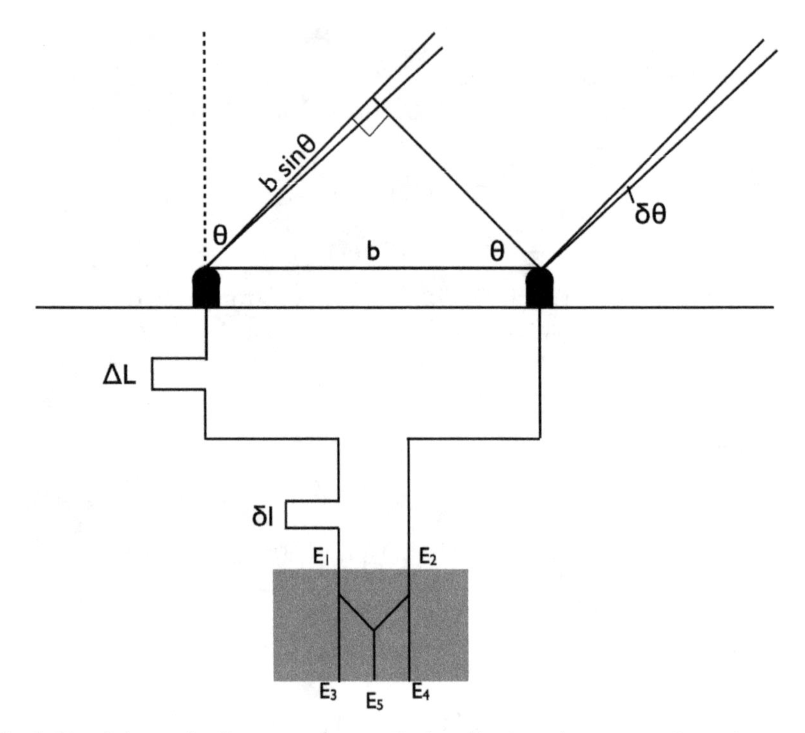

Fig. 3.19. Schematic diagram of an optical stellar interferometer observing a pair of point sources.

but the amplitude of the fringes is a function of the baseline separation and the angular separation of the sources.

The resulting fringe pattern is just the sum of two fringe patterns each given by Equation 3.32, but with zenith angles, θ and θ' differing by the source separation, $\delta\theta = \theta - \theta'$. Note that because the two sources are not in phase, there will be no terms involving the product of both sources, since these will average to zero.

$$\begin{aligned} E_{\text{double}} &= 2fE_0^2\left(2 + \cos\frac{2\pi}{\lambda}(b\sin\theta + \Delta L + \delta l)\right. \\ &\quad \left. + \cos\frac{2\pi}{\lambda}(b\sin\theta' + \Delta L + \delta l)\right) \\ &= 2fE_0^2\left(2 + \cos\frac{2\pi}{\lambda}\delta l + \cos\frac{2\pi}{\lambda}(b\sin\theta' + \Delta L + \delta l)\right). \end{aligned} \tag{3.35}$$

Now, if we assume that θ is small, and setting $\delta l = 0$ for the time being, then

$$E_{\text{double}} \approx 2fE_0^2\left(3 + \cos\left(\frac{2\pi}{\lambda}b\delta\theta\right)\right). \tag{3.36}$$

Therefore, the interference fringes have a maximum contrast when

$$\frac{2\pi}{\lambda}b\delta\theta = m2\pi, \tag{3.37}$$

where m is an integer, and therefore

$$\delta\theta_{\text{double}} = \frac{\lambda}{b}, \tag{3.38}$$

where $\delta\theta_{\text{double}}$ is the minimum angular separation between two point sources which can be resolved.

For an extended source, the expression becomes

$$\delta\theta_{\text{extended}} = 1.22\frac{\lambda}{b}. \tag{3.39}$$

The fringes can be scanned by rapidly varying δl, but the fringe amplitude, or fringe visibility, will be determined by b and $\delta\theta$. The fringe visibility is defined as

$$V_{\text{fringe}} = \frac{I_{\text{max}} - I_{\text{min}}}{I_{\text{max}} + I_{\text{min}}}, \tag{3.40}$$

where I_{max} and I_{min} are the minimum and maximum intensities. By observing the source at different baseline separations, e.g. with an array of telescopes, or at different times throughout the night when the rotation of the earth alters the projected baseline, one obtains a variation in the fringe visibility from which the separation or size of the source may be determined, see, e.g. Figure 3.20.

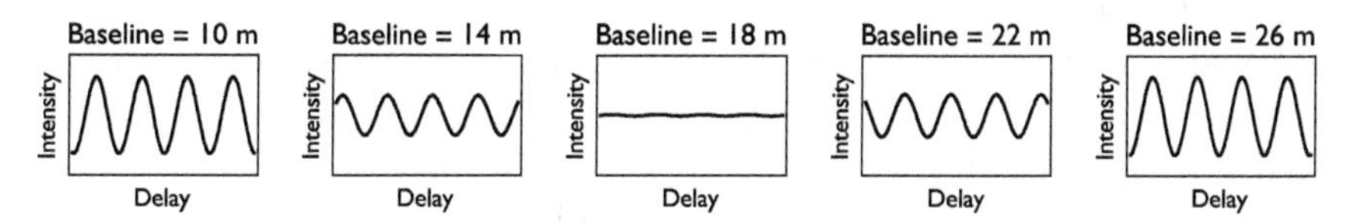

Fig. 3.20. Schematic example of the visibility of the fringes produced by a pair of point sources observed with a two-aperture optical interferometer as a function of baseline. The scanning of the fringes is done by modulating the path length of one arm, and hence the fringe separation is unaffected by the changing baseline.

The aperture size does not affect the angular resolution, which is determined by the separation between apertures, but rather the limiting brightness of the interferometer: a larger aperture collects more light and measurements can therefore be made for fainter sources.

3.5.2 *Beam transport, optical path length matching and delay lines*

It is necessary to transport the light from the telescope apertures to a point at which the beams can be interfered. It is paramount that this is achieved whilst precisely controlling the optical path length. Any unwanted changes in the path length will necessarily change the interference pattern, and therefore defeat the purpose of the interferometer. The optical path length must therefore be controlled to within approximately one wavelength, the exact requirement depending on the bandwidth.

However, as discussed above, it is necessary to include a method of purposely changing the path length of the beams in a controlled manner, both in order to compensate for the sidereal change in zenith angle, and also to incorporate a deliberate modulation of the interference fringes. Again, these deliberate changes of the path length must be controlled to within a wavelength.

3.5.3 *Beam combination*

It is necessary to be able to combine the beams to produce the interference. Within optical interferometry the beams are always added to produce the interference before detection (whereas in radio interferometry the signals can be detected, then multiplied by cross-correlation). There are two main schemes for adding the beams. The first is Fizeau interferometry, in which the beams are added in the image plane, producing a spatially resolved image of fringes at the detector. The second is Michelson interferometry, in which the beams are added in the pupil plane, which therefore does not produce spatial fringes, but temporally resolved fringes at the detector. In this book, we concentrate on Michelson interferometry, since the addition of beams within waveguides, as in astrophotonic applications, is necessarily Michelson interferometry.

3.5.4 *Photometric calibration*

Note, that because the signals in Equation 3.32 are added to produce the interference, there is a term $2E_0^2$, which will be a significant source of noise, containing both Poissonian shot-noise as well as systematically changing due to atmospheric turbulence. It is therefore necessary to calibrate the photometry of the incoming signals accurately to reduce the contribution of this noise.

3.5.5 *Bandwidth*

The interference pattern given by Equation 3.32 is dependent on wavelength, and the examples shown in Figure 3.20 are for monochromatic sources. Increasing the bandwidth will decrease the number of fringes which is approximately $\lambda/\delta\lambda$. Decreasing the bandwidth to obtain a large number of fringes comes at the cost of reduced efficiency, since most photons are then wasted. Typically, a compromise of $\lambda/\delta\lambda \approx 10$ is used, and the path length is swept by tens of microns (Labeyrie *et al.* 2014). If using Fizeau interferometry the interferogram can be dispersed to obtain an image of the fringes as a function of wavelength.

3.5.6 *Aperture masking*

The examples above assume the beams from two separate telescopes are being combined. It is also possible to create interference fringes from a single telescope by applying an aperture mask to the pupil. The mask may be applied at any pupil image within the optical path, e.g. in place of the aperture stop in Figure 2.6(b).

For this technique to work the holes must be smaller than Fried's parameter, r_0, which is the largest diameter across which the light passing through the earth's atmosphere remains in phase (see Section 2.1).

The exposures must also be kept short, such that the random noise added to the phase at each aperture is 'frozen' (as in speckle interferometry). In this case, so long as there are at least three apertures, the atmospheric noise can be removed by obtaining the closure phase. For example, let the random phase due to atmospheric turbulence at each of three apertures at a particular instant be e_1, e_2 and e_3. The measured difference in phase between pairs of apertures, Φ_{ij} is then

$$\Phi_{12} = \phi_{12} + e_1 - e_2,$$

$$\Phi_{23} = \phi_{23} + e_2 - e_3,$$

$$\Phi_{31} = \phi_{31} + e_3 - e_1,$$

where ϕ_{ij} are the true phases due to the path length difference from the source to the aperture. The closure phase is given by

$$\Phi_c = \Phi_{12} + \Phi_{23} + \Phi_{31}, \quad (3.41)$$

which is independent of the atmospheric terms e_i.

Each pair of apertures results in an independent fringe measurement, provided that the separation and orientation of the apertures is unique, i.e. the mask in non-redundant.

The small apertures and short exposure times limit aperture masking to relatively bright sources, but the technique provides excellent resolution at the diffraction limit of the telescope.

CHAPTER 4

PROPAGATION OF LIGHT IN WAVEGUIDES

The preceding two chapters introduced some of the basic concepts of astronomical instrumentation. We now turn our attention to the basic concepts of photonics, beginning with the principles of the propagation of light in waveguides. This is a fundamental background for understanding all of the astrophotonic applications which follow. We will start with a heuristic description of a two-dimensional slab waveguide. This will allow us to develop all the necessary concepts of the propagation of light in waveguides in an accessible manner. In the next chapter, we will extend these concepts to various types of real waveguides used in photonics. For an introduction to waveguide modes, see, e.g. Quimby (2006), for a rigorous treatment and derivations, see, e.g. Snyder and Love (1983) or Okamoto (2006).

4.1 Optical Waveguides

Optical waveguides work on the principle of total internal reflection of light as it propagates down a waveguide. Consider the diagram in Figure 4.1 depicting a two-dimensional slab waveguide. Light is injected from a medium with refractive index n_0 into the core of a waveguide with refractive index n_1, which is surrounded by a cladding with lower refractive index n_2.

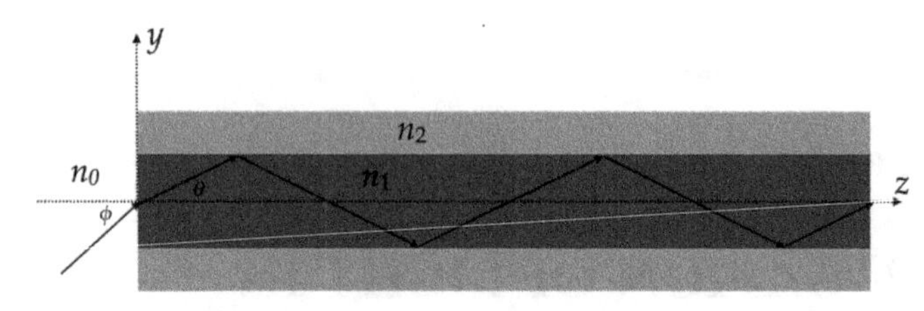

Fig. 4.1. Sketch of the principle of total internal reflection within a slab waveguide, and definition of the refractive index and angles.

The light in the core of the waveguide will be totally internally reflected if

$$n_1 \cos\theta \geq n_2$$
$$\implies n_1 \sin\theta \leq \sqrt{n_1^2 - n_2^2}, \tag{4.1}$$

thus the condition on the injection angle, ϕ, for total internal reflection is

$$n_0 \sin\phi \leq \sqrt{n_1^2 - n_2^2}, \tag{4.2}$$

where the quantity $n_0 \sin\phi$ is called the **numerical aperture** or NA of the waveguide.

Example 4.1: What is the approximate fractional index difference

$$\Delta n = \frac{n_1 - n_2}{n_1},$$

for a silica waveguide with $n_1 = 1.462$ and $NA = 0.11$?
From Equation 4.2, we have

$$NA = \sqrt{n_1^2 - n_2^2}$$
$$\implies NA^2 = n_1^2 - n_2^2$$
$$\implies n_2 = \sqrt{1.462^2 - 0.11^2} = 1.45786,$$

and therefore

$$\Delta n = 2.8 \times 10^{-3}. \tag{4.3}$$

The numerical aperture is related to the input f-ratio (Section 2.2.3), since

$$\tan\phi = \frac{1}{2F}, \tag{4.4}$$

and therefore, if ϕ is small

$$NA = n_0 \sin\phi \approx n_0 \tan\phi \approx \frac{n_0}{2F}. \tag{4.5}$$

Example 4.2: Consider a fibre with a numerical aperture of $NA = 0.22$ being fed in air. What is the equivalent f-ratio, and half-angle, of the input beam?
From Equation 4.5

$$\begin{aligned} F &\approx \frac{n_0}{2NA} \\ &\approx \frac{1}{2 \times 0.22} \\ &\approx 2.3. \end{aligned}$$

The half-angle of the input beam is given by Equation 4.2

$$\begin{aligned} NA &= n_0 \sin\phi, \\ \implies \phi &= \sin^{-1} NA, \\ &\approx 12.7 \text{ degrees}. \end{aligned}$$

A familiar example of an optical waveguide is an optical fibre, in which the core and cladding are made from glass with different refractive indices. Such dielectric waveguides are the most common type of optical waveguides. Non-dielectric waveguides which still operate on the principle of total internal reflection also exist, including waveguides made from semiconductors such as Si, and photonic crystal fibres (Section 5.6). However, there are other types of waveguides which use different phenomena to confine the wave. For example, hollow conducting waveguides, such as those used at microwave wavelengths, confine the wave through reflections off the conducting walls of the waveguides. Another important example is photonic bandgap fibres (which will be discussed in Section 5.6.2), which use a periodic medium to create band gaps at certain wavelengths, allowing the light to be guided. For now, we will consider only those waveguides which operate on the principle of total internal reflection.

4.2 Waveguide Modes

Because the electromagnetic field in a waveguide is confined, there must exist specific modes of the electromagnetic field within the waveguide, as for any confined wave. We will look at the origin of these modes in two ways. First, we will describe the origin of modes by considering a wave propagating along a waveguide, and determining those waves which can propagate indefinitely. Second, we will look at the electromagnetic field distribution in cross-section across the waveguide as determined by Maxwell's equations and the boundary conditions imposed by the waveguide. Following this, we will describe the fundamental properties of waveguide modes.

4.2.1 *The origin of waveguide modes*

Equation 4.2 gives the criterion for total internal reflection in a waveguide. However, not all light rays which satisfy this condition are able to propagate down the waveguide indefinitely. Only those with wave fronts that add together in phase after repeated reflections will continue to propagate.

Consider the light rays in Figure 4.2. At points A and B both rays are on the same wavefront, and are in phase. As the ray propagates from point B to point C the optical path length is (see Example 4.3),

$$\overline{BC} = n_1 \frac{2a}{\sin\theta}\left(\cos^2\theta - \sin^2\theta\right), \tag{4.6}$$

where a is the radius of the waveguide. Assuming that the light has wavelength λ_0 in vacuum, then the phase change over this path length is

$$\Phi_{\mathrm{BC}} = \frac{2\pi}{\lambda_0} n_1 \frac{2a}{\sin\theta}\left(\cos^2\theta - \sin^2\theta\right). \tag{4.7}$$

Now consider the ray which travels from point A to D, which starts out in phase the ray at point B. This time the phase change is,

$$\Phi_{\mathrm{AD}} = \frac{2\pi}{\lambda_0} n_1 \frac{2a}{\sin\theta} + 2\Phi_r, \tag{4.8}$$

where Φ_r is the phase change induced at each reflection.

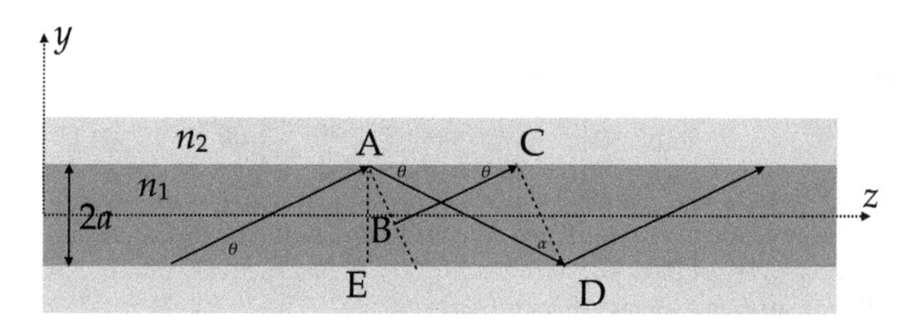

Fig. 4.2. Sketch of a light ray propagating down a waveguide.

Example 4.3: Prove Equation 4.6.
Consider the triangle ABC. This gives us

$$\cos\theta = \frac{\overline{BC}}{\overline{AC}}. \tag{4.9}$$

Meanwhile, from triangle ADE, we have

$$\sin\theta = \frac{2a}{\overline{AD}}. \tag{4.10}$$

And from triangle ACD, the law of sines gives us

$$\begin{aligned}
\frac{\sin\alpha}{\overline{AC}} &= \frac{\sin\left(\theta + \frac{\pi}{2}\right)}{\overline{AD}} \\
\frac{\sin\left(\frac{\pi}{2} - 2\theta\right)}{\overline{AC}} &= \frac{\cos\theta}{\overline{AD}} \\
\frac{\cos 2\theta}{\overline{AC}} &= \frac{\cos\theta}{\overline{AD}} \\
\frac{\cos^2\theta - \sin^2\theta}{\overline{AC}} &= \frac{\cos\theta}{\overline{AD}}.
\end{aligned} \tag{4.11}$$

Then substituting Equations 4.9 and 4.10 into 4.11 gives,

$$\begin{aligned}
\frac{\cos\theta\left(\cos^2\theta - \sin^2\theta\right)}{\overline{BC}} &= \frac{\sin\theta\cos\theta}{2a} \\
\overline{BC} &= \frac{2a}{\sin\theta}\left(\cos^2\theta - \sin^2\theta\right),
\end{aligned} \tag{4.12}$$

which multiplied by the refractive index gives the optical path length in Equation 4.6.

If the wave at C is to remain in phase with the wave at D, then the phase difference between the two rays must be a multiple integer of 2π, i.e.

$$\Phi_{\mathrm{AD}} - \Phi_{\mathrm{BC}} = \frac{2\pi}{\lambda_0} n_1 4a \sin\theta + 2\Phi_r = 2\pi m. \tag{4.13}$$

If this condition is satisfied the wave will constructively interfere with itself as it propagates down the waveguide, and will thus propagate indefinitely. Waves propagating at other angles will destructively interfere, and will not survive over significant distances.

The phase shift at each reflection, Φ_r, is dependent on the polarisation of the light. If the electric field is parallel to the boundary between the core and the cladding, then there is no component in the z direction, and the light is said to be transverse electric, or **TE**, polarised. Conversely, if the magnetic field is parallel to the boundary between the core and the cladding, the light is said to be transverse magnetic, or **TM**, polarised. The phase shift for TE and TM modes (the Goos–Hänchen shift) is given by

$$\Phi_r = \begin{cases} 2\tan^{-1}\left(\dfrac{\sqrt{\cos^2\theta - \left(\dfrac{n_2}{n_1}\right)^2}}{\sin\theta}\right), & \text{TE}, \\ 2\tan^{-1}\left(-\left(\frac{n_1}{n_2}\right)^2 \dfrac{\sqrt{\cos^2\theta - \left(\frac{n_2}{n_1}\right)^2}}{\sin\theta}\right), & \text{TM}. \end{cases} \tag{4.14}$$

Finally, then, the condition for a light wave propagating down a waveguide to remain in phase with itself is given by

$$\tan\left(\frac{2\pi}{\lambda_0} n_1 a \sin\theta - \frac{m\pi}{2}\right) = \begin{cases} \dfrac{\sqrt{\cos^2\theta - \left(\dfrac{n_2}{n_1}\right)^2}}{\sin\theta}, & \text{TE}, \\ -\left(\dfrac{n_1}{n_2}\right)^2 \dfrac{\sqrt{\cos^2\theta - \left(\dfrac{n_2}{n_1}\right)^2}}{\sin\theta}, & \text{TM}. \end{cases} \tag{4.15}$$

Equation 4.15 cannot be solved analytically, but must be solved numerically, or graphically, see, e.g. Figure 4.3. The electromagnetic field distributions which satisfy Equation 4.15 for each m are the modes of the waveguide. Figure 4.4 shows some examples of different modes in a slab

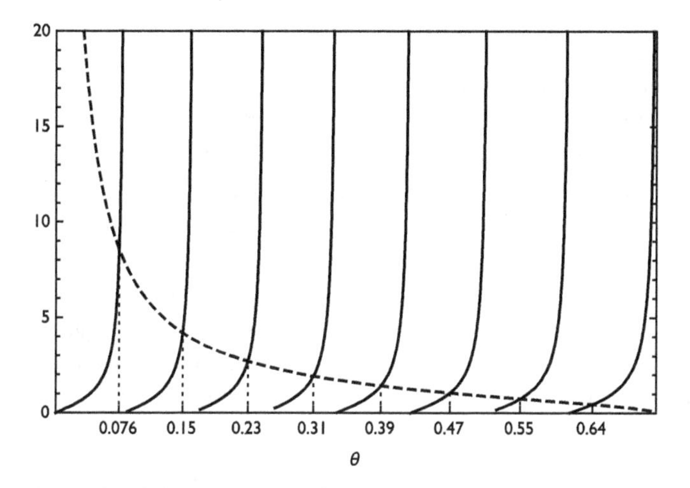

Fig. 4.3. Example of the graphical solution of Equation 4.15 for TE modes, with the LHS shown by the black lines, and the RHS shown by the dotted line, for $n_1 = 1.9$, $n_2 = 1.44$, $a = 5$ μm and $\lambda = 1.55$ μm. There are eight modes which have the propagation angles denoted on the x-axis.

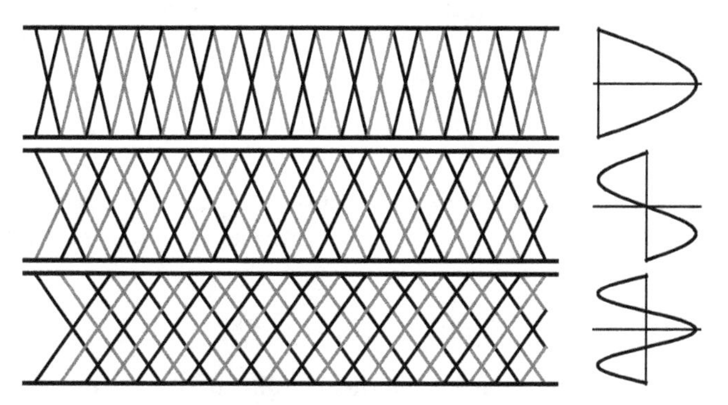

Fig. 4.4. Schematic depiction of the formation of modes by the constructive interference of reflected wavefronts propagating down a waveguide (after Okamoto 2006). When the wave peaks (black) and troughs (grey) add together in phase the electromagnetic field distribution is stable (right hand plots), and the light can propagate down the waveguide.

waveguide. Thus, the electromagnetic field within a waveguide is a superposition of the various allowable modes, which are not generally in phase, and do not generally have the same amplitude.

4.2.2 The electromagnetic field distribution of modes

We will now consider the origin of modes by considering the electromagnetic field distribution across the waveguide at any point and at any time. The electromagnetic field distribution of each mode within the waveguide can in principle be found by solving Maxwell's equations, with the boundary conditions that the field must decay exponentially outside the waveguide. In practice, this is only possible analytically for situations with simplifying assumptions, such as specific geometries, and with certain approximations, e.g. for specific polarisations. Considering again the two-dimensional slab waveguide, the solution to Maxwell's equations for each mode within the waveguide must be a wave equation,

$$\nabla^2 E(\mathbf{r},t) - \frac{1}{c^2}\frac{\partial^2 E(\mathbf{r},t)}{\partial t^2} = 0. \tag{4.16}$$

Separating the time dependent and position dependent parts of the electric field,

$$E(\mathbf{r},t) = E(\mathbf{r})\mathrm{e}^{\mathrm{i}\omega t}, \tag{4.17}$$

the time independent part of the wave equation then becomes

$$\nabla^2 E(\mathbf{r}) - k^2 E(\mathbf{r}) = 0, \tag{4.18}$$

which is known as the **Helmholtz equation**, and $k = \omega/c = 2\pi/\lambda$ is the **wavenumber**.

Now, returning to our slab waveguide the time independent electric field can be further separated into a z component, and a transverse component

$$E(y,z) = a_m E_m(y)\mathrm{e}^{-\mathrm{i}\beta_m z}, \tag{4.19}$$

where a_m are constants which give the amplitude of each mode.

The z dependent term is simply a sinusoidal variation with z, with propagation constant β_m, which gives the phase change per unit length along the z-axis, i.e.

$$\beta_m = k_0 n_1 \cos\theta_m, \tag{4.20}$$

where $k_0 = 2\pi/\lambda$ is the wavenumber (in vacuum) and θ_m is the angle of the mth mode as given by Equation 4.15.

Since, we are dealing with a simplified two-dimensional waveguide, the transverse component only depends on y. This term can be determined by substituting Equation 4.19 into the Helmholtz Equation, and requiring that the field internal to the waveguide is continuous with the external field at the core-cladding boundary. Doing so, yields

$$E_m \propto \begin{cases} e^{-\gamma_m y}, & y > \frac{a}{2}, \\ e^{\gamma_m y}, & y < -\frac{a}{2}, \\ \cos\left(2\pi \frac{\sin\theta_m}{\lambda} y\right), & m = 0, 2, 4, \ldots \text{ and } -\frac{a}{2} \leq y \leq \frac{a}{2}, \\ \sin\left(2\pi \frac{\sin\theta_m}{\lambda} y\right), & m = 1, 3, 5, \ldots \text{ and } -\frac{a}{2} \leq y \leq \frac{a}{2}, \end{cases} \tag{4.21}$$

where γ_m is the extinction coefficient,

$$\gamma_m = \sqrt{\beta_m^2 - n_2^2 k_0^2}. \tag{4.22}$$

Thus, each mode is described by a sinusoidally varying electric field across the waveguide, and an exponentially decaying field in the cladding. The constants of proportionality in Equation 4.21 can be determined by ensuring that u_m^2 is normalised, i.e.

$$\int_{-\infty}^{\infty} E_m^2(y)\, \mathrm{d}y = 1. \tag{4.23}$$

Figure 4.5 shows an example for the modes in the same slab waveguide of Figure 4.3.

4.2.3 *Properties of modes*

Before going on to discuss the modes of more realistic waveguides in Chapter 5, it is worth highlighting a few important properties of modes.

Evanescent field. The field distribution of each mode extends into the cladding. This is known as the evanescent field and is a very important property for the function of many photonic devices, since it enables

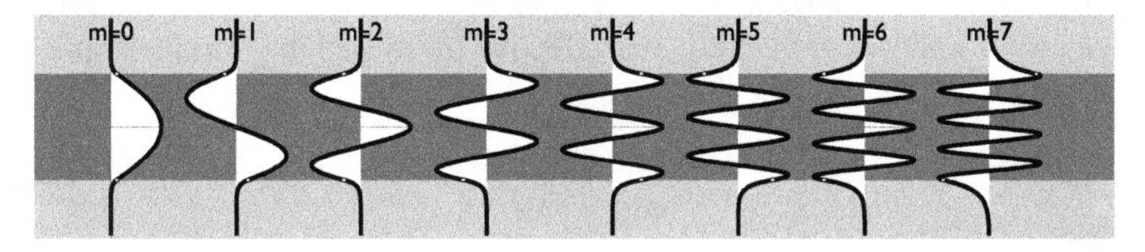

Fig. 4.5. Example of the electric field distribution for a slab waveguide with $n_1 = 1.9$, $n_2 = 1.44$, $a = 5$ μm and $\lambda = 1.55$ μm, which has eight modes with propagation angles as shown in Figure 4.3.

the field from one waveguide to couple to an adjacent waveguide (Section 6.5.2). Note that the evanescent field is only evident from the electromagnetic treatment of waveguide; this is one of many cases for which ray optics is not sufficient to describe photonic devices.

Effective index and propagation constant. Because the field extends into the cladding, the propagation of each mode is not determined purely by the material index of the core, but also partly by the cladding, such that the effective index, n_{eff}

$$n_2 < n_{\mathrm{eff}} < n_1. \tag{4.24}$$

Furthermore, because each mode has a different electric field distribution (e.g. Figure 4.5), each mode will have a different effective index and a different propagation constant. This is an important caveat for the consideration of multimode photonics (Chapter 12).

The propagation constant (Equation 4.20) is the component of the wavenumber in the material of the core in the z-direction, i.e. along the axis of the waveguide, and it is related to the effective index by

$$\beta_m = k_0 n_1 \cos\theta_m = k_0 n_{\mathrm{eff}_m}. \tag{4.25}$$

Phase velocity. The phase velocity of the waveguide is the rate of change of phase in the z-direction, i.e.

$$v_{\mathrm{p}} = \frac{\omega}{\beta_m} = \frac{c}{n_{\mathrm{eff}_m}}. \tag{4.26}$$

Group velocity. The group velocity is the velocity at which a pulse would move down the waveguide, i.e.

$$v_{\mathrm{g}} = \frac{c}{n_1}\cos\theta_m = \frac{\mathrm{d}\omega}{\mathrm{d}\beta}. \tag{4.27}$$

Group index. Analogously to the relation between velocity and refractive index, and phase velocity and effective index, we may define a group index,

$$v_{\mathrm{g}} = \frac{c}{n_{\mathrm{g}_m}}, \tag{4.28}$$

which implies

$$n_{\mathrm{g}_m} = \frac{n_1}{\cos\theta_m}. \tag{4.29}$$

Note also that

$$v_g = \frac{d\omega}{d\beta} = \frac{d\omega}{d\lambda}\frac{d\lambda}{d\beta}. \tag{4.30}$$

Now, since $\omega = 2\pi c/\lambda$, we have

$$\frac{d\omega}{d\lambda} = -\frac{2\pi c}{\lambda^2}. \tag{4.31}$$

And as $\beta = 2\pi n_{\text{eff}}(\lambda)/\lambda$, and in general the refractive index is a function of λ,

$$\frac{d\beta}{d\lambda} = -\frac{2\pi n_{\text{eff}}(\lambda)}{\lambda^2} + \frac{2\pi}{\lambda}\frac{dn_{\text{eff}}}{d\lambda}. \tag{4.32}$$

Therefore,

$$n_g = n_{\text{eff}}(\lambda) - \lambda\frac{dn_{\text{eff}}}{d\lambda}. \tag{4.33}$$

Polarisation. Consider a wave propagating in the z-direction in a slab waveguide, as in the top panel of Figure 4.6. In this case, both the electric field (grey) and the magnetic field (black) are transverse to the propagation direction, and the wave is said to be transverse electromagnetic (TEM).

Now, consider a wave propagating down a slab waveguide in the yz-plane, but at an angle to the z-axis. Let the y and z components of the electric field $E_y = E_z = 0$, as in the middle panel of Figure 4.6. Therefore, the electric field is transverse to the z direction and the wave is said to be transverse electric, or TE polarised.

Likewise, if the x and z components of the electric field $E_x = E_z = 0$, as in the bottom panel of Figure 4.6, then the magnetic field is transverse to the z direction and the wave is said to be transverse magnetic, or TM polarised.

4.2.4 *Dispersion*

Consider a pulse of light injected into a waveguide. As the light travels along the waveguide the pulse will gradually become more spread out. This spreading is called **dispersion**. Dispersion can arise from several causes.

4.2.4.1 *Modal dispersion*

In a multimode waveguide each mode propagates with a different mode angle (Equation 4.15), and therefore travels with a different group velocity, as given in Equation 4.27. Therefore, each mode of a pulse in a multimode waveguide will arrive at a specific point in the waveguide at a different time due to the different path lengths taken. Graded index fibres are designed to minimise such dispersion, and will be discussed in Section 5.2.

4.2.4.2 *Material dispersion*

In a single-mode fibre monochromatic light would suffer no dispersion. However for light with a spread in wavelength, the dependence of the material refractive index on wavelength and the variation of the effective index with wavelength both cause **chromatic dispersion**.

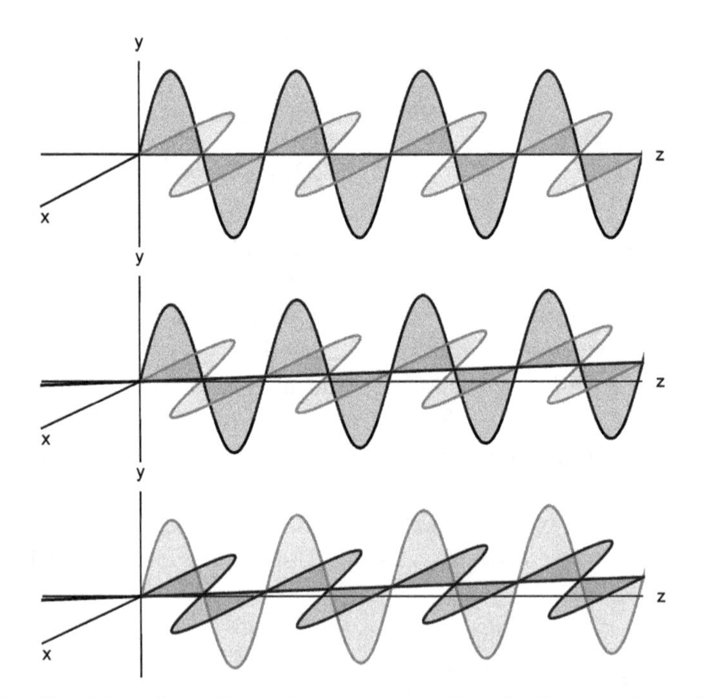

Fig. 4.6. Possible polarisations of a wave travelling in the yz plane of a slab waveguide. In the top panel, both electric (grey) and magnetic (black) fields are transverse, and the wave is TEM polarised. In the middle panel, only the electric field is transverse, and the wave is TE polarised. In the bottom plot the magnetic field is transverse and the wave is TM polarised.

The dispersion due to the dependence of the material refractive index on wavelength is known as **group velocity dispersion** or **material dispersion**.

Consider a pulse injected at $z = 0$, $t = 0$. The time taken to travel a distance Z is

$$\begin{aligned} t &= \frac{Z}{v_g} \\ &= \frac{Z}{c}\left(n_{\text{eff}}(\lambda) - \lambda\frac{\mathrm{d}n_{\text{eff}}}{\mathrm{d}\lambda}\right). \end{aligned} \tag{4.34}$$

Thus, if the pulse has a spread in wavelengths of $\Delta\lambda$ the spread in travel time will be

$$\begin{aligned} \Delta t &= \frac{\mathrm{d}t}{\mathrm{d}\lambda}\Delta\lambda \\ &= D_\lambda Z\Delta\lambda, \end{aligned} \tag{4.35}$$

where

$$D_\lambda = -\frac{\lambda}{c}\frac{\mathrm{d}^2 n_{\text{eff}}}{\mathrm{d}\lambda^2}, \tag{4.36}$$

is the dispersion coefficient.

4.2.4.3 *Waveguide dispersion*

The second cause of chromatic dispersion in single-mode fibres is the different mode profiles for different wavelengths in the same waveguide. The extinction coefficient, γ_m, in Equation 4.21 determines how much the electric field of any mode extends into the cladding, and is directly related to the propagation constant, and therefore the effective index via Equations 4.22 and 4.20. The group velocity also depends on n_{eff} via Equations 4.28 and 4.33. Now, even if the material refractive index were constant with wavelength, n_{eff} would still be dependent on the wavelength due to the different electric field distribution of different wavelengths. The propagation constant is directly related to n_{eff} and the mode angle, θ_m, via Equation 4.20. However, θ_m depends on λ, as well as the refractive indices of the core and cladding and the waveguide radius from Equation 4.15. Thus, n_{eff}, and by extension the group velocity, depends upon the wavelength, giving rise to **waveguide dispersion**. This term derives from the fact that the dispersion is determined by the structure of the waveguide.

4.2.5 *Focal ratio degradation*

An optical fibre has a maximum acceptance angle given by its *NA* (Equation 4.2). Unless light is fed at exactly this maximum angle it will exit the fibre at a wider angle (faster focal ratio) than the injected light. This phenomenon is called **focal ratio degradation (FRD)**, since the focal ratio is degraded in the sense that the étendue (Section 2.4) is increased. In the language of photonics FRD is sometimes known as ***NA* up-scattering**. Figure 4.7 shows examples of FRD for a typical fibre and a good fibre.

For a fibre feeding a spectrograph (one of the most common uses of fibres in astronomy) FRD is always bad, even if it is well characterised and compensated for. This is because FRD will cause either a decrease in throughput (e.g. if the FRD is not allowed for in subsequent optics), a loss

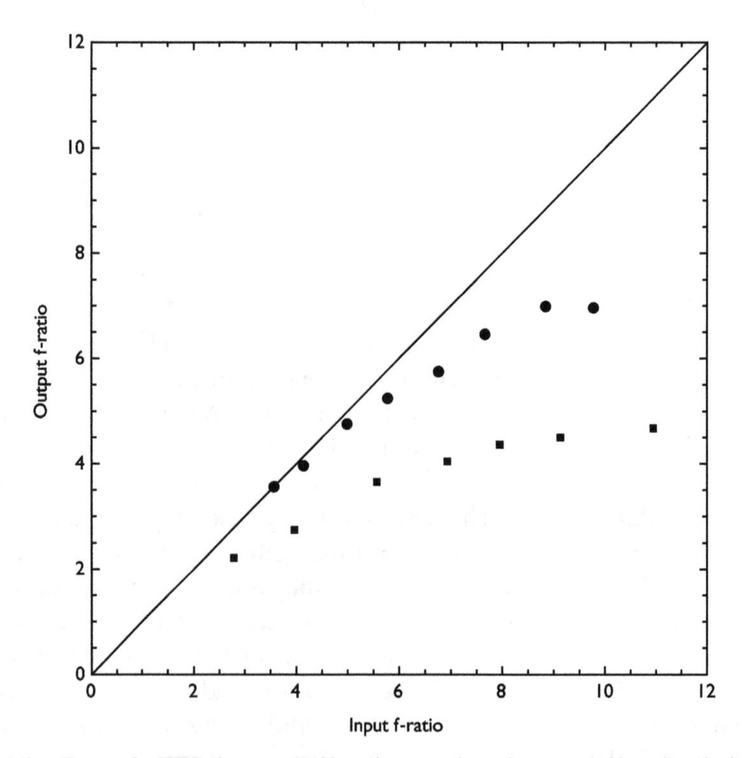

Fig. 4.7. Example FRD for a test fibre (squares) and a good fibre (circles) from Carrasco and Parry (1994). Data for the good fibre are from Barden (1987). The straight line shows the relation for no FRD.

of resolving power (e.g. if FRD is compensated for in the spectrograph entrance slit), a loss of multiplex gain, or a loss of wavelength coverage (since the light is more spread out, each spectral resolution element requires more pixels). Furthermore, compensating for FRD with larger and faster optics increases the cost of the spectrograph. Therefore, FRD is to be avoided if possible by feeding the fibre at a fast speed close to its natural *NA*.

CHAPTER 5

WAVEGUIDES IN PRACTICE

The general properties of waveguides were discussed in Chapter 4 from a simplified treatment of a two-dimensional slab waveguide. We now consider different types of real waveguides as used in photonics, and discuss their properties.

5.1 Step-Index Fibres

The most common type of waveguide used in astronomy is the **step-index fibre**. This is the familiar fibre with a cylindrical core of constant refractive index, surrounded by a concentric cladding of lower constant refractive index. This is usually coated in a protective polymer buffer, e.g. acrylate or polyimide. The fractional refractive index difference between the core and the cladding, $\Delta = (n_1 - n_2)/n_1$ is usually small, with typical values of 10^{-2}–10^{-3}.

5.1.1 *Fibre materials*

The fibre material is chosen for the particular wavelength range required. For astronomical applications the most common choices are fused silica with either high OH content for good UV throughput, or low OH content for good NIR throughput, or a compromised broadband fibre. Fused silica fibre becomes opaque above ≈ 1.8 μm due to phonon–phonon absorption in the glass, which has traditionally limited the use of fibre optics in astronomy to wavelengths shorter than this. However, there are alternative materials, such as chalcogenide glasses, and fluoride glasses (e.g. ZBLAN),

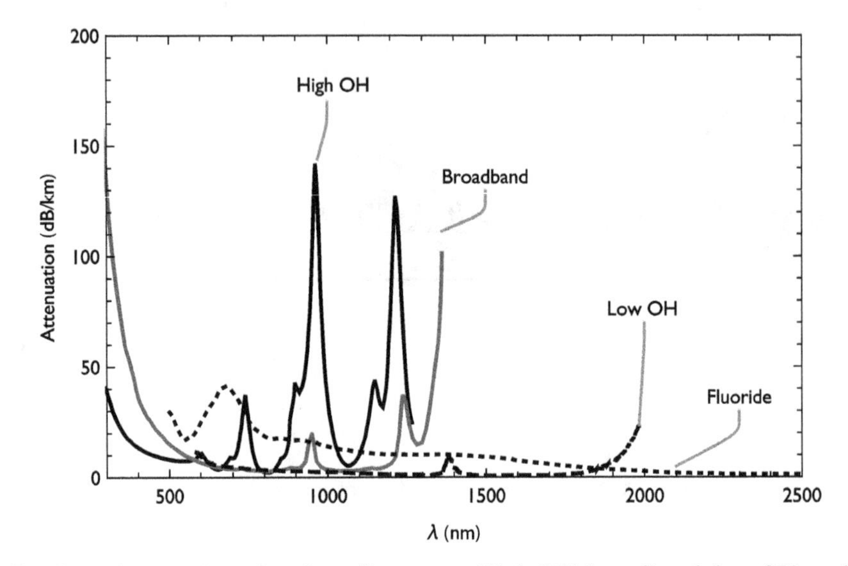

Fig. 5.1. Attenuation of various fibre types: High OH, broadband, low OH, and ZrF_4.

which have good transmission throughout the NIR (Haynes *et al.* 2006). Unfortunately, these glasses tend to be more difficult to handle, as both are fragile and fluoride glass is also hygroscopic, while chalcogenide often contains toxic materials such as As or Sb. Figure 5.1 shows measured attenuation profiles for various fibre types.

5.1.2 *Modes of a step-index fibre*

The modes in a step-index fibre can be described as a mixture of radial and azimuthal modes. For example, in the case of fibres with a small refractive index contrast (known as 'weakly guiding' fibres, because the confinement of the field to the core is not so strong[1]), the propagation of the light down the fibre is close to the optical axis of the fibre, and therefore the electromagnetic fields are approximately transverse, with negligible components in the longitudinal direction. In this case, the strict description of modes as either TE (transverse electric) or TM (transverse magnetic) modes can be replaced with an approximate description of linearly polarised (LP)

[1]N.B. The term 'weakly guiding' *does not* imply that the fibres are lossy, or that the fibres do not guide well.

modes. Then the electric field may be written as

$$E(r, \phi, z) = E_{lm}(r) \cos l\phi \, e^{i(\omega t - \beta z)}, \tag{5.1}$$

where

$$E_{lm}(r) \propto \begin{cases} J_l(k_{T_m} r), & r \leq a, \\ K_l(\gamma_m r), & r > a, \end{cases} \tag{5.2}$$

where J_l is a Bessel function of the first kind, and K_l is a modified Bessel function, and

$$k_T^2 = n_1^2 k_0^2 - \beta^2, \tag{5.3}$$

$$\gamma^2 = \beta^2 - n_2^2 k_0^2, \tag{5.4}$$

which can be found by solving the characteristic equation (i.e. the equivalent of solving Equation 4.15),

$$X \frac{J_{l+1}(X)}{J_l(X)} = \pm Y \frac{K_{l+1}(Y)}{K_l(Y)}, \tag{5.5}$$

$$X^2 + Y^2 = V^2, \tag{5.6}$$

where

$$X = k_T a, \tag{5.7}$$

$$Y = \gamma a, \tag{5.8}$$

and V is the normalised frequency,

$$V = 2\pi \frac{a}{\lambda_0} \sqrt{n_1^2 - n_2^2}, \tag{5.9}$$

and a is the fibre radius. There are exactly equivalent modes with an orthogonal polarisation given by Equation 5.1 with a $\sin l\phi$ azimuthal dependence. The constants of proportionality for Equation 5.2 can be found be ensuring that $E_{lm}(r)$ is continuous at a, and the overall normalisation can be found be ensuring $\int_0^\infty E_{lm}(r)\, dr = 1$.

There can be multiple solutions, $m = 1, 2, 3, \ldots$ to the characteristic equation for each l. Thus, each mode is denoted by the l and m mode numbers, where m denotes the number of intensity peaks in radius, and $2l$ describes the number of intensity peaks in azimuth. Furthermore, for $l \neq 0$, there are always two solutions to the characteristic Equation (5.5) with $\pm l$ corresponding to the plus or minus forms of the equation. These two solutions have the same roots, and the same propagation constants etc., but

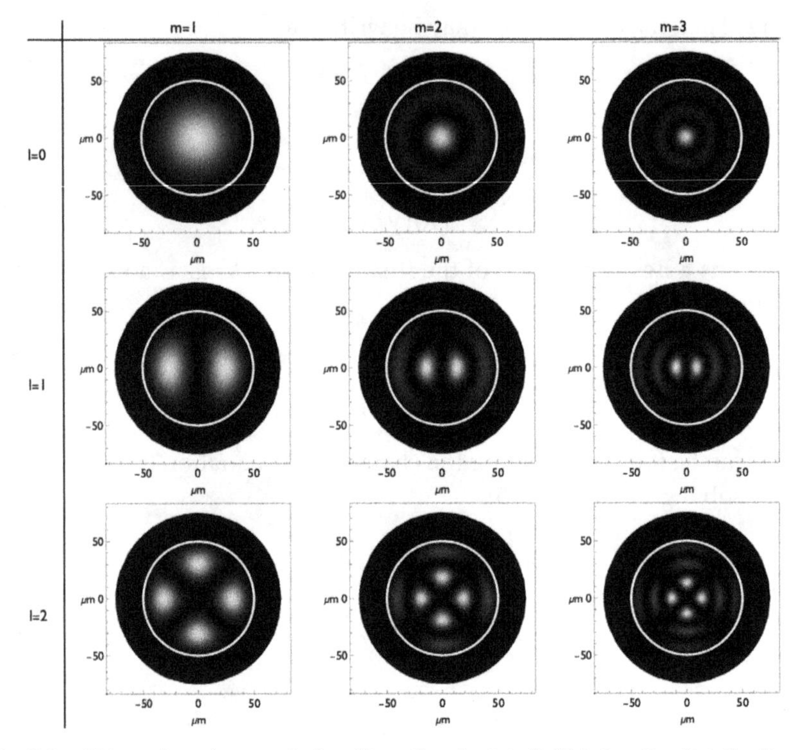

Fig. 5.2. LP modes of a step-index fibre: the electric field intensity distribution of the first nine modes in a fibre with $a = 50\ \mu\text{m}$, $n_1 = 1.51$, $n_2 = 1.5$ at $\lambda_0 = 1.55\ \mu\text{m}$. The white circle marks the circumference of the fibre core. N.B. for $l \neq 0$ there are always two identical modes, and each mode can exist in one of two polarisations.

correspond to helical modes propagating with equal but opposite helicity; modes with $l = 0$ correspond to meridional rays. Figure 5.2 shows an example for the first nine modes in a fibre with $a = 50\ \mu\text{m}$, $n_1 = 1.51$, $n_2 = 1.5$ at $\lambda_0 = 1.55\ \mu\text{m}$.

The number of modes in an optical fibre can be approximated by (Senior 1992)

$$M = \frac{V^2}{4}, \tag{5.10}$$

where we have ignored differences in polarisation, since in the general case these are unimportant in astronomical applications. Another commonly used approximation (Saleh and Teich 2007) is

$$M \approx \frac{1}{2}\left(2 + \frac{V^2}{\pi^2}\right). \tag{5.11}$$

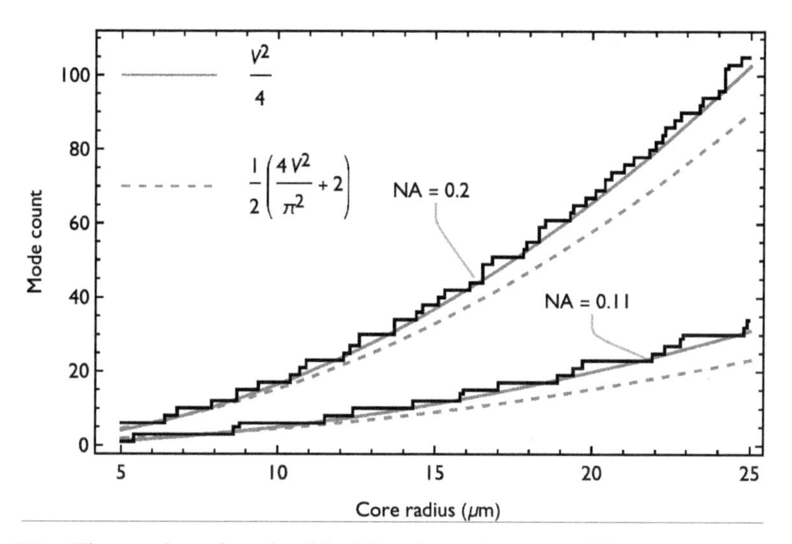

Fig. 5.3. The number of modes (black lines) as a function of fibre core radius for a step-index fibre at a wavelength of 1.55 μm compared to the approximation given by Equations 5.10 and 5.11 (grey lines).

Figure 5.3 compares these approximations to the full calculation of modes using Equations 5.2–5.9, for two example fibres. The former approximation is superior when $M > 9$, and we prefer this for few-mode fibres. However, both approximations fail for low mode counts.

Example 5.1: Calculate the maximum core diameter for a single-mode fibre with $NA = 0.11$ at a wavelength of 0.5 μm and 1.5 μm.
From Equations 5.9 and 5.12, we have

$$2\pi\frac{a}{\lambda_0}NA \leq 2.405$$

$$\implies a \leq \frac{2.405\lambda_0}{2\pi NA}.$$

Therefore, for $\lambda_0 = 0.5$ μm we have $2a \leq 3.5$ μm, and at $\lambda_0 = 1.5$ μm we have $2a \leq 10.4$ μm.

5.1.2.1 *Single-mode fibre*

From the full calculation of modes, when

$$V \leq 2.405, \tag{5.12}$$

the fibre can carry only one mode, and is referred to as a **single-mode fibre (SMF)**.

The output electric field of a single-mode fibre can be approximated by a Gaussian beam described by

$$E_{\mathrm{SMF}} = E_0 \mathrm{e}^{-\frac{x^2}{w^2}}, \tag{5.13}$$

and therefore the intensity is given by

$$I = I_0 \mathrm{e}^{-\frac{2x^2}{w^2}}. \tag{5.14}$$

The point at which the intensity drops to $1/\mathrm{e}^2$ of the maximum is given by,

$$x = \pm w, \tag{5.15}$$

and is known as the **beam waist**, so the **mode field diameter** is given by $d_{\mathrm{smf}} = 2w$.

When the normalised frequency (Equation 5.9), $1.2 < V < 2.4$, the mode waist can be approximated as

$$w \approx a\left(0.65 + \frac{1.619}{V^{1.5}} + \frac{2.879}{V^6}\right), \tag{5.16}$$

where a is the core radius.

The distinction between a single-mode fibre and a multimode fibre is very important for astrophotonics, since many astrophotonic devices can only work in single-mode waveguides. As we will see, many photonic devices rely on the constructive or destructive interference of separate waves within the same waveguide, and this can only be accomplished if the phases of the waves are well defined at all points. However, for multimode waveguides, note that for each mode LP_{lm}, there is a unique solution to the characteristic equation yielding k_{T} and γ, and therefore each mode has a unique propagation constant and a unique effective index. Therefore, different modes will not remain in phase, even if their optical path lengths are the same. Thus, addition or cancellation of separate waves cannot be properly accomplished in multimode waveguides.

5.2 Graded Index Fibres

In a **graded index fibre (GRIN)**, the refractive index of the core is not constant, but decreases with radius. Therefore, the guided rays do not propagate along the fibre in straight line segments, but follow a curved path. A major application of graded index fibres is minimising the dispersion due to the pulse spreading of different modes in a multimode fibre (see Section 4.2.4.1). In a step-index fibre each mode travels at a different angle along the fibre, and therefore has a different optical path length. Therefore, over a given distance of propagation, each mode will arrive at different times.

In a GRIN fibre, the modes with larger angles still travel further physically; however, they travel faster as they spend more time in a lower refractive index part of the fibre; the optical path lengths can thus be matched by choosing an appropriate refractive index profile. In particular, the intermodal dispersion can be minimised if the refractive index profile is quadratic, e.g.

$$n(r)^2 = n_1^2\left(1 - \frac{\delta^2}{a^2}r^2\right), \tag{5.17}$$

$$\delta^2 = \frac{n_1^2 - n_2^2}{n_1^2}, \tag{5.18}$$

where a is the fibre radius, and $r^2 = x^2 + y^2$. Assuming that $n_1 - n_2 \ll 1$, this refractive index profile results in a sinusoidal path,

$$x(z) = \frac{a\theta_{x0}}{\delta}\sin\frac{\delta}{a}z, \tag{5.19}$$

$$y(z) = y_0\cos\frac{\delta}{a}z + \frac{a\theta_{y0}}{\delta}\sin\frac{\delta}{a}z, \tag{5.20}$$

where y_0 is the y value at the input face, and θ_{x0} and θ_{y0} are the initial injection angles in the x and y planes, and we have set $x_0 = 0$ without loss of generality. Figure 5.4 shows an example of a path along a GRIN fibre, which is exaggerated to show the helicity.

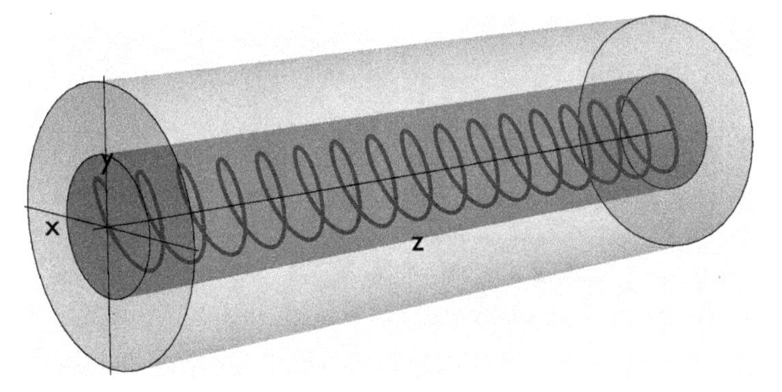

Fig. 5.4. Example of the helical path of a ray in a graded index fibre. The image is not shown to scale, in order to make clear the helicity.

5.3 Fibre Bragg Gratings

We now consider a special class of fibre in which the refractive index of the core varies along the direction of the optical axis, i.e. in the direction in which the light is propagating.

Consider the fibre sketched in Figure 5.5, which depicts a simplified fibre Bragg grating with a core consisting of a periodic array of M high refractive index (n_1) layers with a pitch Λ, embedded in low refractive index (n_0) background, where each layer is exactly $\Lambda/2$ wide. Let light with electric field amplitude E_0 be injected from the left hand side, and travel in the positive z direction.

As the light propagates from the n_0 material to the n_1 material it will be partially reflected, with a reflection coefficient given by Fresnel's equations, and likewise for light propagating from n_1 to n_0,

$$\sqrt{R_{01}} = \frac{n_0 - n_1}{n_0 + n_1} = \sqrt{R}, \tag{5.21}$$

$$\sqrt{R_{10}} = \frac{n_1 - n_0}{n_0 + n_1} = -\sqrt{R}, \tag{5.22}$$

where the difference in sign is due to the π phase change introduced when light is reflected from a medium of higher refractive index.

Consider light reflected from the first n_1 layer. The amplitude of the light reflected from the front face is

$$E_{1_{\text{front}}} = E_0\sqrt{R}. \tag{5.23}$$

The amplitude from the back face is the same, except for a change of sign on the reflection coefficient, and a phase delay $\phi/2$ where ϕ is the

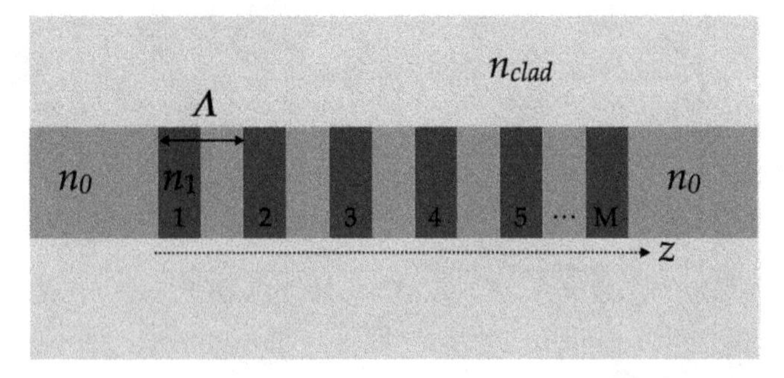

Fig. 5.5. A simplified fibre Bragg grating, with a core having alternate high (n_1) and low (n_0) refractive index material, with a pitch Λ.

phase difference associated with the round-trip distance between two front faces, i.e.

$$E_{1_{\text{back}}} = -E_0\sqrt{R}\mathrm{e}^{\mathrm{i}\frac{\phi}{2}}, \tag{5.24}$$

$$\phi = \frac{2\pi}{\lambda_0}2n\Lambda, \tag{5.25}$$

where $n \approx n_1 \approx n_2$ and we have assumed that $\sqrt{R} \ll 1$. The amplitudes of the light reflected from the second layer are the same (ignoring multiple reflections since $\sqrt{R} \ll 1$), except there is a further phase delay of ϕ, i.e.

$$E_{2_{\text{front}}} = E_0\sqrt{R}\mathrm{e}^{\mathrm{i}\phi}, \tag{5.26}$$

$$E_{2_{\text{back}}} = -E_0\sqrt{R}\mathrm{e}^{\mathrm{i}\left(\phi+\frac{\phi}{2}\right)}. \tag{5.27}$$

Thus, the amplitude of light reflected from the mth layer is

$$E_{m_{\text{front}}} = E_0\sqrt{R}\mathrm{e}^{\mathrm{i}(m-1)\phi}, \tag{5.28}$$

$$E_{m_{\text{back}}} = -E_0\sqrt{R}\mathrm{e}^{\mathrm{i}\left((m-1)\phi+\frac{\phi}{2}\right)}, \tag{5.29}$$

or

$$\begin{aligned} E_m &= E_{m_{\text{front}}} + E_{m_{\text{back}}} \\ &= E_0\sqrt{R}\mathrm{e}^{\mathrm{i}(m-1)\phi}\left(1 - \mathrm{e}^{\mathrm{i}\frac{\phi}{2}}\right). \end{aligned} \tag{5.30}$$

The amplitude of the total reflected light from M layers is then

$$\begin{aligned} E_{\mathrm{r}} &= E_0\sqrt{R}\left(1 - \mathrm{e}^{\mathrm{i}\frac{\phi}{2}}\right)\sum_{m=1}^{M}\mathrm{e}^{\mathrm{i}(m-1)\phi} \\ &= E_0\sqrt{R}\left(1 - \mathrm{e}^{\mathrm{i}\frac{\phi}{2}}\right)\left(\frac{\mathrm{e}^{\mathrm{i}M\phi} - 1}{\mathrm{e}^{\mathrm{i}\phi} - 1}\right). \end{aligned} \tag{5.31}$$

The total reflected intensity is therefore

$$\begin{aligned} I_{\mathrm{r}} &= |E_{\mathrm{r}}|^2 \\ &= I_0 R\frac{\sin^2\frac{M\phi}{2}}{\cos^2\frac{\phi}{4}}. \end{aligned} \tag{5.32}$$

This is a maximum when $\phi = 2\pi, 6\pi, 10\pi, \ldots$ such that the reflected light from every layer adds together in phase, which occurs if the incident wavelength is

$$m\lambda_{\rm B} = 2n\Lambda, \tag{5.33}$$

where m is an integer. The phase difference between the front and back faces is $\phi/2 = \pi$ which would make waves out of phase, except this is compensated by the change of sign in the reflection coefficient; the total phase difference between light reflected from the front and back faces is 2π.

This type of periodic structure is known as a **Bragg grating** due to its analogy with Bragg diffraction in a crystal, and $\lambda_{\rm B}$ is the **Bragg wavelength**, and a fibre with this structure in the core is a **fibre Bragg grating** or **FBG**. FBGs play an important part in astrophotonics and will be discussed in Chapter 8.

At the Bragg wavelength the total reflectivity is

$$\begin{aligned} R_{\rm max_{weak}} &= 4RM^2, \\ &= (\kappa L)^2, \end{aligned} \tag{5.34}$$

where $L = M\Lambda$ is the length of the grating, and $\kappa = \frac{2\Delta n}{\lambda_{\rm B}}$. Equation 5.34 is only true in the case of a weak grating, for which $\sqrt{R} \ll 1$, and therefore $\kappa L \ll 1$. More generally for a strong grating, we have

$$R_{\rm max} = \tanh^2 \kappa L. \tag{5.35}$$

The exact form for κ depends on the refractive index profile, e.g. if the variation is sinusoidal rather than the step profile discussed above

$$\kappa = \frac{\pi\Delta n}{\lambda_{\rm B}}. \tag{5.36}$$

Furthermore, κ gives the attenuation coefficient of the electric field at resonance, i.e.

$$E(z) = E_0 {\rm e}^{-\kappa z}. \tag{5.37}$$

The reflectivity drops to zero at $2\pi \pm 2\pi/M$, i.e. $\Delta\phi = 2\pi/M$, where

$$\begin{aligned} \Delta\phi &= -\frac{2\pi}{\lambda_0^2} 2n\Lambda\Delta\lambda, \\ \frac{2\pi}{M} &= -\frac{2\pi}{\lambda_0^2}\lambda_{\rm B}\Delta\lambda, \\ \frac{\lambda_0}{\Delta\lambda} &\approx M = \frac{L}{\Lambda}, \end{aligned} \tag{5.38}$$

where in the last step we have used $\lambda_{\rm B} \approx \lambda_0$. In the language of astronomical instrumentation the quantity $\lambda_0/\Delta\lambda$ is referred to as the resolving power of the Bragg grating; in photonics it is referred to as the **quality**, or simply Q. In any case it is given simply by the number of grating periods.

A periodic FBG will have a series of Bragg wavelengths for different integer m in Equation 5.33. The spacing between each Bragg wavelength is called the **free spectral range**. Differentiating Equation 5.33 gives,

$$\frac{\mathrm{d}m}{\mathrm{d}\lambda} = -\frac{2n\Lambda}{\lambda^2}, \tag{5.39}$$

and setting $\mathrm{d}m = 1$ yields the free spectral range,

$$\delta\lambda = \frac{\lambda^2}{2n\Lambda}. \tag{5.40}$$

A simple sinusoidal variation in refractive index over a finite length of fibre will produce a reflection such as that shown in Figure 5.6(a). The characteristics of the FBG are as described above, but there are also secondary peaks in reflection due to the sudden discontinuity between a constant index and a varying index. These secondary peaks can be suppressed with a gradual transition between the constant and varying parts of the

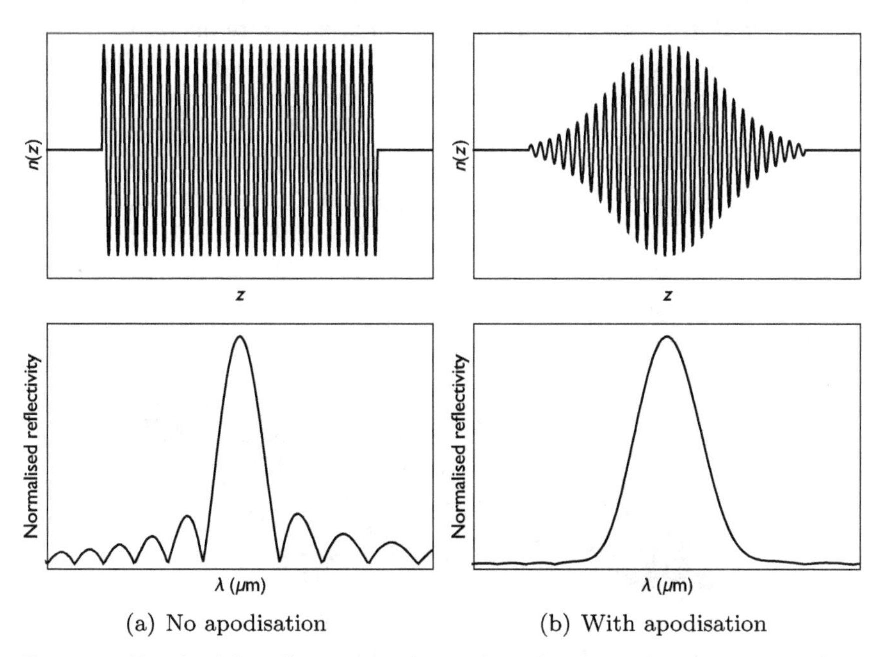

(a) No apodisation (b) With apodisation

Fig. 5.6. Sketch of the effects of apodising the refractive index variations, which serves to reduce the secondary reflections in the Bragg grating.

index, such as the variation shown in Figure 5.6(b). This pattern of index variation is called an **apodised** FBG.

More generally, the reflectivity of any index variation can be found by taking the Fourier transform of the index profile, $R(k) = \mathcal{F}(n(z))$, and then making the substitution $k \rightarrow \frac{4\pi n_0}{\lambda}$. This process can also be reversed to find the required refractive index profile for a specific response.

5.4 Channel Waveguides

A very important class of waveguides in photonics is that of **channel waveguides**. These are essentially rectangular cross-section waveguides, the exact geometry of which may vary due to the manufacturing process. Examples are shown in Figure 5.7. The reason channel waveguides are important in photonics is that they can be lithographically printed onto a wafer to form a two-dimensional **photonic integrated circuit** (**PIC**), incorporating many types of photonic components, see, e.g. Figure 1.2.

It is possible to make channel waveguides using many different types of materials. Important examples are Ti:$LiNbO_3$, in which a channel of Ti is diffused into a $LiNbO_3$ substrate. **Silicon-on-insulator** (**SOI**, sometimes also referred to as silica-on-silicon) uses a Si substrate on which a layer of SiO_2 is deposited, on top of which is another thin layer of Si. This final layer of Si is etched away to leave a Si ridge waveguide on top of the lower index SiO_2. The ridge waveguide may be clad with a low index material such as SiO_2 or PMMA, or left in air. SOI is a very active area of research for photonics, since the manufacture can be done using CMOS fabrication technology, and the high index contrast between Si and SiO_2 (or PMMA or air) allows very tight bending radius waveguides to be used without serious bending losses. However, the high index contrast also means that the waveguides have very small dimensions, and are therefore difficult to

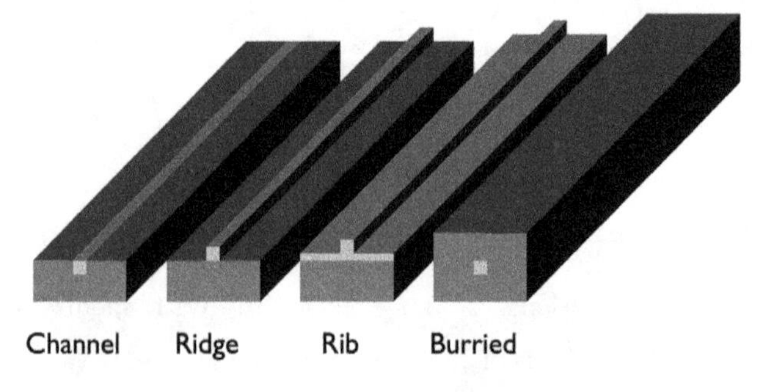

Fig. 5.7. Sketch of the geometries of example channel waveguides.

couple into (see Chapter 6). A compromise is to use other CMOS compatible materials such as Si_3N_4, or SiO_xN_{1-x} in place of the Si waveguide layer.

5.4.1 *Modes of rectangular waveguides*

5.4.1.1 *Polarisations*

Recall the possible polarisations of a slab waveguide, as in Figure 4.6. Because the wave vector was confined to the yz plane it was possible for the propagating wave to have purely transverse electric or magnetic components, or both.

For a real three-dimensional waveguide the wave will not in general be confined to a plane, and thus it is not possible to ensure that $E_z = H_z = 0$, and strictly speaking we cannot have TE, nor TM modes. However, there are modes in which either the E_x and H_y components are dominant, or alternatively modes in which the E_y and H_x components are dominant. These are referred to as the E^x and E^y modes respectively.

The E^x and E^y modes will have different mode profiles due to the fact that the waveguide is not symmetric. Therefore, the different polarisations will have different effective indices , and will not propagate in phase with one another. This birefringence is an important consideration in the practical application of channel waveguides, especially those with high refractive index contrast.

5.4.1.2 *Electric field distribution*

From our previous introductory discussion of modes in Section 4.2.2, we anticipate that the electric field distribution in a rectangular waveguide will be of the form

$$E \propto E(x)E(y),$$

$$E(x) = \begin{cases} \cos m\dfrac{\pi x}{a}, & m = 1, 3, 5, \ldots \\[2ex] \sin m\dfrac{\pi x}{a}, & m = 2, 4, 6, \ldots \end{cases} \tag{5.41}$$

$$E(y) = \begin{cases} \cos p\dfrac{\pi y}{d}, & p = 1, 3, 5, \ldots \\[2ex] \sin p\dfrac{\pi y}{d}, & p = 2, 4, 6, \ldots \end{cases} \tag{5.42}$$

within the waveguide core, i.e $-\frac{a}{2} \leq x \leq \frac{a}{2}$ and $-\frac{d}{2} \leq y \leq \frac{d}{2}$, and to be exponentially decaying outside the core, and indeed this is the case.

It is not possible to solve Maxwell's equations for the general case of a wave in a rectangular waveguide. However, there are some useful approximate methods.

Marcatili's method (Marcatili 1969; Okamoto 2006), assumes that the waveguide can be approximated by the refractive index profile shown in Figure 5.8, i.e. that the field in the corners is negligible. It then proceeds by assuming that $H_x = 0$ for the E^x modes, or conversely that $H_y = 0$ for the E^y modes. Figure 5.9 shows the E_x distribution for the E^x modes of an example waveguide. This example, which has $n_1 = 1.99$, $n_2 = 1.44$, highlights the fact that the assumption that the corner fields are negligible is not appropriate for waveguides with a high index contrast — there is a significant fraction of the field distribution missing.

There are a number of extensions to Marcatili's method appropriate for different approximations. Westerveld *et al.* (2012) provide an extension for high index contrast waveguides. The assumption that $H_x = 0$ for the E^x

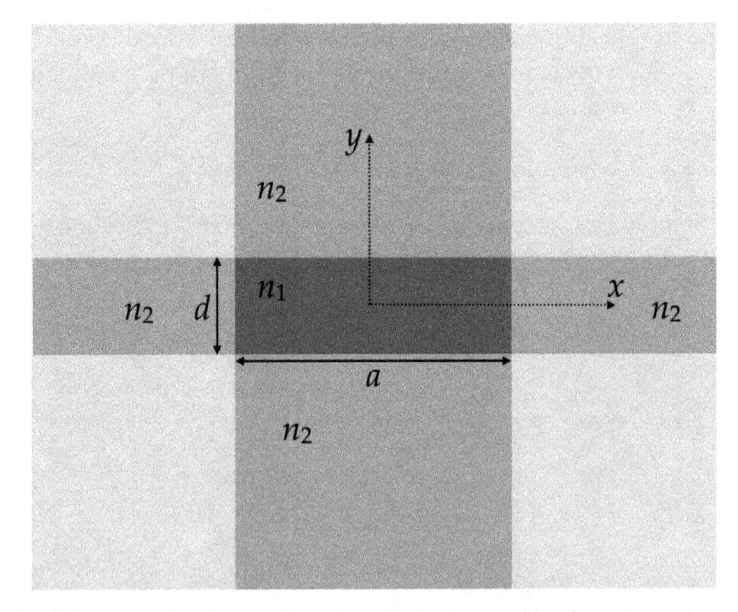

Fig. 5.8. The approximation of the refractive index profile of a rectangular waveguide in Marcatili's method. The field in the corner regions is assumed to be negligible.

modes (or $H_y = 0$ for the E^y modes) is replaced with the requirement that the tangential electromagnetic fields are continuous across the boundaries in the x direction, while the E_x field is explicitly matched across the waveguide boundaries in the y direction, see the second panel of Figure 5.9. However, the refractive index profile is the same as for Marcatili's method (Figure 5.8), and so again assumes that the modes are well confined.

Kumar's method (Kumar *et al.* 1983; Okamoto 2006), approximates the refractive index profile in the corner regions as $\approx \sqrt{2n_2^2 - n_1^2}$, and so overcomes this problem, but this approximation can only be valid for waveguides with low index contrast such that $2n_2^2 > n_1^2$.

In general, it is more practical to use numerical methods, such as the beam propagation method, to find the modes of realistic waveguides. There are many commercially available software packages available for mode solving. In Figure 5.9, we compare results using `RSOFT BeamPROP` from `Synopsys` to the Marcatili method, for E^x modes.

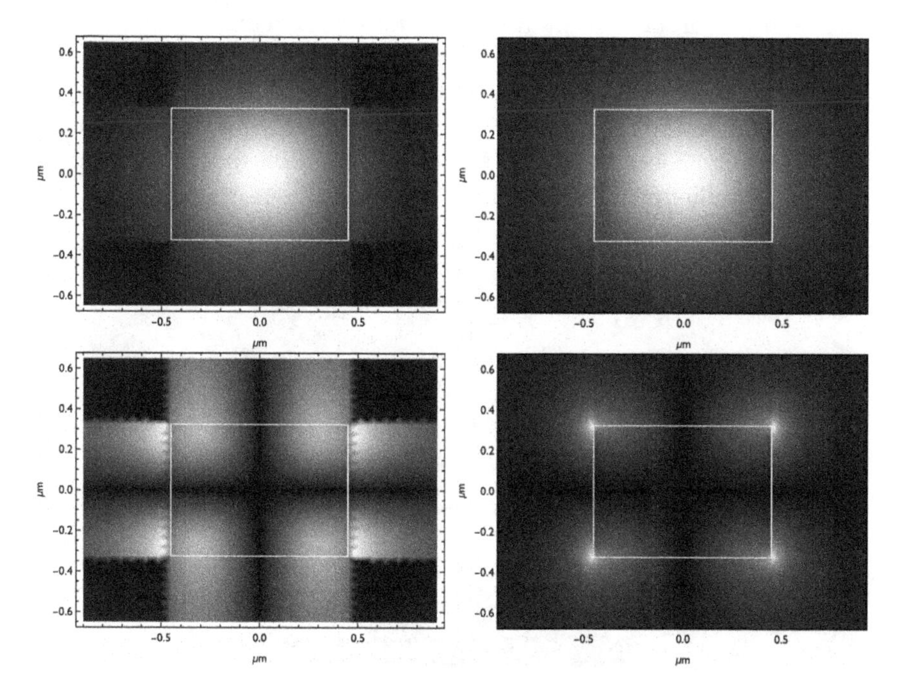

Fig. 5.9. The E^x modes of a waveguide with width = 0.9 μm, height = 0.65 μm, $n_1 = 1.99$, $n_2 = 1.44$ and $\lambda = 1.55$ μm. The left hand panels shows E_x calculated using Marcatili's method, and the right hand panel shows the results from `RSOFT BeamPROP`.

5.5 Direct-Write Waveguides

Direct-write waveguides are inscribed into a solid block of glass. Therefore the layout of the waveguides can be precisely arranged and remains fixed, as for channel waveguides. Unlike channel waveguides however, they are not confined to a plane, but can be arranged in three dimensions. This extra versatility allows them to be used for applications which would be difficult or impossible to implement otherwise (see, e.g. Cai and Wang 2022).

Direct-write waveguides operate on the same principle of total internal reflection to the fibres and channel waveguides discussed above. The core of the waveguide is written by focussing a high power laser beam (typically a femto-second laser) into the glass (see, e.g. Davis *et al.* 1996; Farsari *et al.* 2019). The power from the laser modifies the structure of the glass, and creates a spot of higher refractive index. The glass block can then be translated under computer control to write a waveguide, see Figure 5.10. The geometry of the waveguides is generally not circular in cross-section, but for single-mode waveguides the fundamental mode is nevertheless nearly symmetric and Gaussian.

The refractive index contrast is typically $\sim 10^{-3}$ for direct-write waveguides, similar to optical fibres. Moreover, the refractive index contrast can be controlled through the laser power, the writing speed, and by

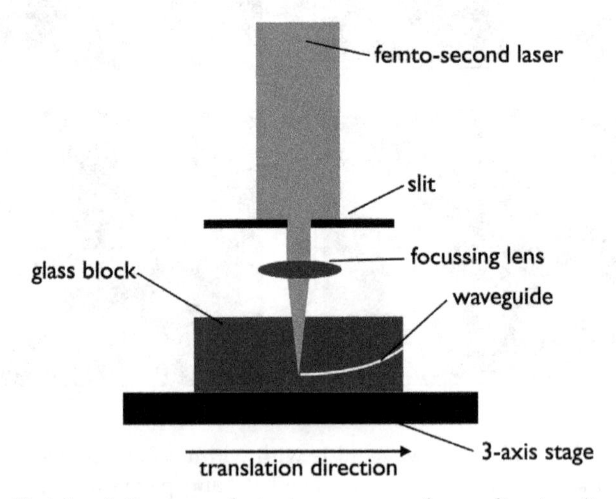

Fig. 5.10. Sketch of the manufacturing process for a direct-write waveguide written with a femto-second laser (see Ams *et al.* 2008).

multiple passes. This allows periodic structures such as Bragg gratings (see Section 5.6.2) to be written in the waveguide, either during the writing process, or after. Unlike lithographic processes, very high index contrasts are not possible, which places practical limits on the bend radii of waveguides.

5.6 Photonic Crystal Waveguides

All the examples discussed above operate on the principle of total internal reflection, which is achieved by surrounding a core of higher index material with a cladding of lower index material. We now examine two important classes of waveguides for which this is not the case. In the first case, the light is again confined to the core by total internal reflection, but this is achieved by constructing a cladding from the same material of the core, but which is punctuated with an array of holes. In the second case, the light is confined to the core not by total internal reflection, but because the structure of the waveguide forbids the propagation of light in the cladding. As we shall see, these photonic crystal structures allow some remarkable behaviours which are not possible with the normal core-cladding types of waveguides.

5.6.1 *Photonic crystal fibres*

A **photonic crystal fibre** consists of hexagonally packed rods and tubes of the same material (e.g. silica), arranged as shown in the left hand panel of Figure 5.11. This preform is then drawn into a fibre, as a conventional fibre would be. The central tube is replaced with a rod, such that the resultant fibre consists of a solid core, surrounded by a cladding of an array of air holes in the same material, as in the right hand panel of Figure 5.11.

The fibre guides light via total internal reflection, with the index difference provided by the air holes which lower the average index of the cladding. Nevertheless, a PCF has some important differences, both practical and functional, from conventional step-index fibre. From a manufacturing point of view, the fibre can be constructed from a single type of glass. There is no requirement for doping to alter the index of the core.

However, the real interest in PCF is that they exhibit behaviour which is not possible to achieve with step-index fibre. The most important being that they can be made single-mode over a very wide wavelength range.

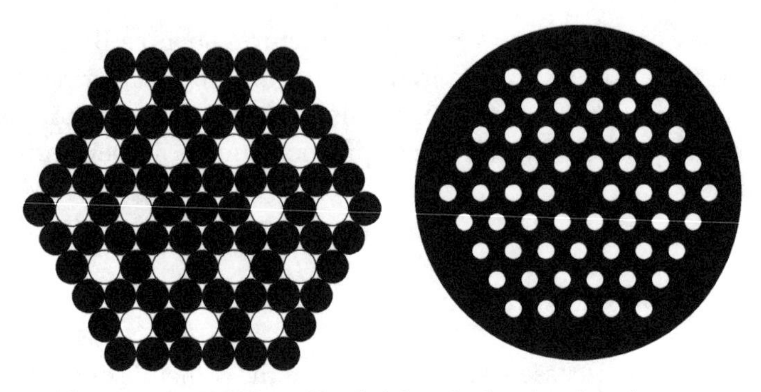

Fig. 5.11. Photonic crystal fibre. The left hand plot sketches the arrangement of solid rods (black discs) and hollow tubes (circles) in a fibre preform. The right hand figure shows a sketch of the resultant PCF. Note the central hole of the fibre is replaced with a solid rod.

Such fibres are referred to as being **endlessly single-mode** (see Birks *et al.* 1997).

Recall that a step-index fibre is single-mode (Equation 5.12, ignoring polarisations) when

$$V \leq 2.405. \tag{5.43}$$

Therefore, substituting Equation 5.9 a particular fibre is only single-mode for wavelengths

$$\lambda_0 \geq 2\pi \frac{a}{2.405}\sqrt{n_1^2 - n_2^2}. \tag{5.44}$$

Now, for a PCF the normalised frequency is given by

$$V_{\mathrm{PCF}} = 2\pi \frac{\Lambda}{\lambda_0}\sqrt{n_1^2 - n_{\mathrm{pcf}}^2}, \tag{5.45}$$

where Λ is the hole spacing, which is the equivalent of the core radius, and n_{pcf} is the effective index for the cladding of a PCF. At long wavelengths the cladding index is given by

$$\lim_{\frac{\Lambda}{\lambda}\to 0} n_{\mathrm{pcf}}^2 = (1-f)n_1^2 + f n_{\mathrm{a}}^2, \tag{5.46}$$

where f is the filling factor of the holes, and n_{a} is the refractive index of the holes, e.g. $n_{\mathrm{a}} = 1$ for air. The filling factor for holes with diameter d is simply their area divided by the area of a hexagon, with a flat-to-flat width of Λ, i.e.

$$f = \frac{\pi}{2\sqrt{3}}\left(\frac{d}{\Lambda}\right)^3. \tag{5.47}$$

At short wavelengths the light is more confined to the glass regions and avoids the holes, such that as $\frac{\Lambda}{\lambda} \to \infty$, $V_{\mathrm{PCF}} \to V_{\mathrm{lim}}$, where V_{lim} is a constant. Therefore

$$\lim_{\frac{\Lambda}{\lambda}\to\infty} n_{\mathrm{pcf}} \approx n_1. \tag{5.48}$$

V_{lim} itself depends on the ratio of the hole size to the pitch, d/Λ. If the glass between the holes gets too thin, then the light cannot be contained in the glass as λ decreases, but is diffracted, and n_{pcf} never reaches n_1. However, if d/Λ is sufficiently small, then $V_{\mathrm{lim}} < 2.405$ for all values of λ. That is, the fibre is always single-mode, no matter how large the core size, or how small the wavelength, provided d/Λ is small enough. This is illustrated in Figure 5.12; for a silica fibre with air holes, and a single missing air hole providing the core, if $d/\Lambda < 0.4$ the fibre is always single-mode. For larger values of d/Λ the fibre is only single-mode for sufficiently large λ. If the core is made from three or seven missing air holes then the limiting values of d/Λ become 0.25 and 0.15, respectively.

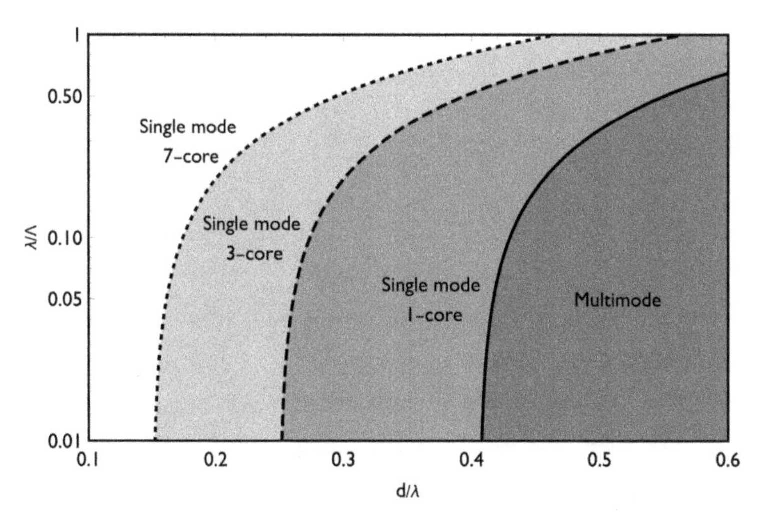

Fig. 5.12. The threshold between single-mode and multimode behaviour for a PCF.

Since the criterion for single-mode behaviour relies only on the ratio d/Λ, and not on d itself, it is possible to construct **large mode area** single-mode fibre, which is not possible for step-index fibre.

5.6.2 *Photonic bandgap waveguides*

Photonic bandgap fibres have a similar periodic structure to photonic crystal fibres, but do not operate by total internal reflection. Rather the structure of the cladding prevents light of certain wavelengths propagating into the cladding due to Bragg reflections.

To understand this phenomenon recall the description of fibre Bragg gratings given in Section 5.3. Light of wavelength $\lambda_B \pm \Delta\lambda$ cannot propagate far along the FBG, and the electric field amplitude will be attenuated by a factor $e^{-\kappa z}$ after a distance z. This is the principle upon which photonic bandgap fibres and waveguides operate.

For example, consider a two-dimensional waveguide such as that shown in Figure 5.13, where the cladding now consists of a Bragg grating. Light at the wavelengths $\lambda_B \pm \Delta\lambda$ cannot propagate into the cladding, due to the Bragg reflections at such wavelengths, and will therefore be

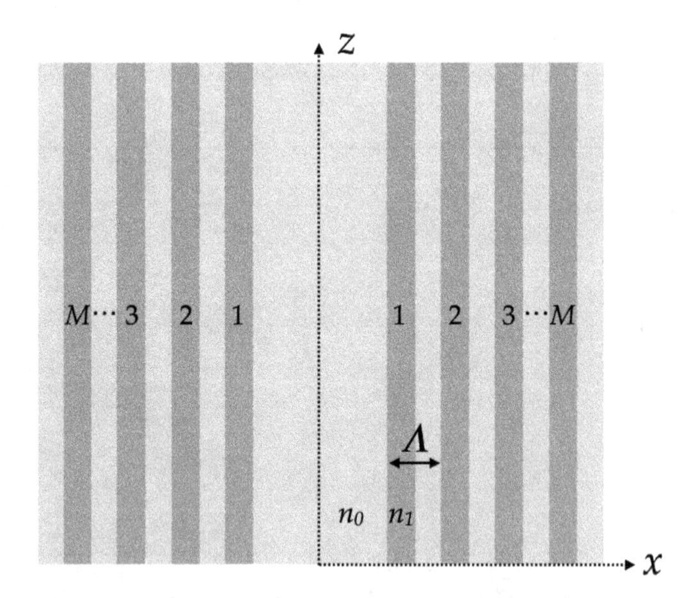

Fig. 5.13. Schematic diagram of a two-dimensional photonic bandgap waveguide. Light at $\lambda_B \pm \Delta\lambda$ cannot propagate into the cladding in the x direction, and is thus confined to the propagate in the z direction within the core.

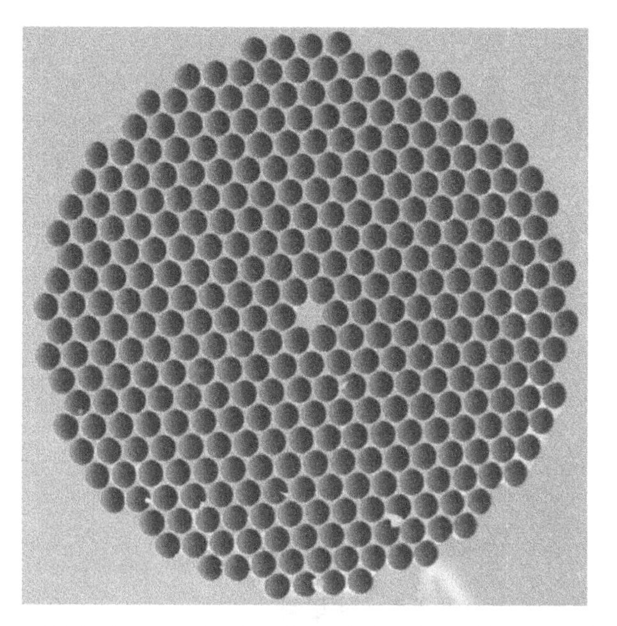

Fig. 5.14. An SEM photograph of a photonic bandgap fibre from the US Naval Research Laboratory. Courtesy of Wikipedia, Public Domain.

confined to the core of the waveguide, and is only able to propagate in the z direction.

These forbidden wavelength ranges are the bandgaps which give photonic bandgap waveguides their name. It is common in the photonics literature to see these described in terms of frequency $\omega = 2\pi/\lambda$ rather than wavelength as we have done here, but the language of wavelength is more natural for astronomy.

Although conceptually easy to understand in terms of one or two-dimensional waveguides, manufacturing practical three-dimensional waveguides is considerably more complicated. Nevertheless, there are now many different types of photonic bandgap fibres available, using different constructions, and with different behaviour, such as that shown in Figure 5.14 (see, e.g. Russell 2003 for further details).

CHAPTER 6

COUPLING LIGHT INTO WAVEGUIDES

The previous chapters have introduced the basic principles of astronomical optics and instrumentation (Chapters 2 and 3), and the basic concepts and properties of waveguides (Chapters 4 and 5). These principles are the foundations upon which astrophotonics is built. We now begin the development of astrophotonics proper by bringing these two strands together. In particular, we look at the crucial issue of coupling of light into waveguides from a telescope, and the reciprocal problem of coupling light out of waveguides into an instrument, as well as coupling between waveguides. We begin with a general consideration of coupling light into a waveguide.

6.1 Coupling into a Waveguide

Consider light being injected into a waveguide from another waveguide or from a lens or telescope, etc. The efficiency of this coupling depends on how well the electric field distribution of the input light overlaps with the electric field distribution of each mode of the waveguide. Specifically the coupling efficiency is given by the **overlap integral**

$$\eta = \sum_{m=1}^{M} \frac{\left| \iint E_0 E_m^* \,\mathrm{d}x\,\mathrm{d}y \right|^2}{\iint |E_0|^2 \,\mathrm{d}x\,\mathrm{d}y \iint |E_m|^2 \,\mathrm{d}x\,\mathrm{d}y}, \tag{6.1}$$

where E_0 is the input electric field, E_m^* is the complex conjugate of the electric field distribution of the mth mode of the waveguide, and there are

M modes in total. The denominator ensures that the efficiency is normalised such that $0 \leq \eta \leq 1$.

Figure 6.1 shows some examples for Gaussian electric fields with varying width and offsets; any difference between the two fields decreases the coupling efficiency.

Specific cases of coupling into waveguides will be given in Section 6.2.2. First, we will examine general schemes for coupling a telescope into a waveguide, and address some of the relevant design options. Thereafter, we will look at the efficiency with which the coupling can be achieved, first in the single-mode and few mode regimes and thereafter in the multimode regime. This will lead to a discussion on adaptive optics, whereby the effects of atmospheric turbulence are partially corrected to reproduce nearly diffraction limited images, allowing for higher coupling efficiencies. Following this we will introduce the photonic lantern, a cornerstone of astrophotonics which allows the efficient conversion of a multimode waveguide into an array of single-mode waveguides and vice versa. Finally, we will examine coupling between waveguides, including from single-mode fibres into channel waveguides, and evanescent coupling between adjacent waveguides.

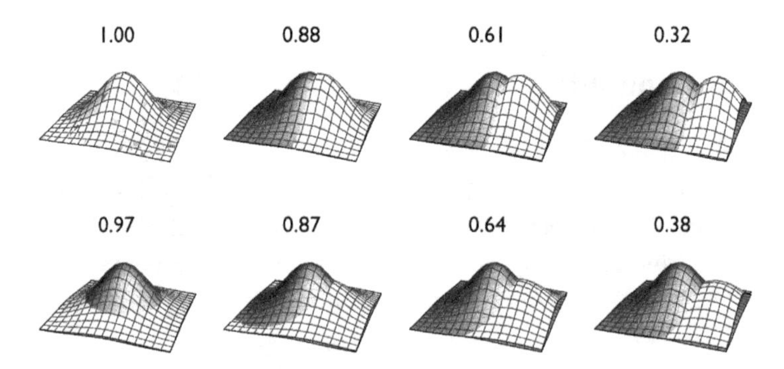

Fig. 6.1. Examples of the coupling efficiency of two Gaussian electric field distributions. The columns show an increasing offset of 0, 0.5, 1 and 1.5 times the σ of the Gaussian between the two fields. In the top row both Gaussians are the same width, in the bottom row the second Gaussian 1.2 times wider than the first.

6.2 Coupling from a Telescope to a Waveguide: General Schemes

All professional optical and near-infrared observational astronomy requires the use of a telescope. The light from the telescope must therefore be coupled into the fibres or waveguides which comprise the astrophotonic instrument. We begin with a general consideration of how to couple light from a telescope to a waveguide, after which we will discuss the efficiency of the coupling. See also the review article by Ellis *et al.* (2021) for further discussion of the results in this section.

6.2.1 *Coupling in the image plane and the pupil plane*

Consider the sketch shown in Figure 6.2(a). Here a fibre is placed directly in the focal plane of the telescope and an image of a star (or galaxy, etc.) is formed on the front face of the fibre. This is coupling in the image plane, and the coupling efficiency would be given by Equation 6.1, with the input electric field given by the profile of the object, e.g. the electric field profile applicable to a Moffat function (Equation 2.3; Figure 2.1) in the case of star observed in natural seeing conditions.

Now, consider the situation in Figure 6.2(b). Here the image of the star is formed on a microlens which is glued to the front face of the fibre. The length of the microlens is equal to its focal length. Therefore, at the back face of the lens, at the entrance to the fibre, an image of the telescope mirror is formed, i.e. a pupil image, and this plane is known as the pupil plane. N.B., in both cases light from the entire telescope mirror forms the image injected into the fibre.

Multiple fibres, with or without microlenses, can be placed on the image plane to capture the light from different parts of the sky simultaneously, e.g. from multiple different stars, as discussed in Section 3.3.3.

Both coupling schemes depicted in Figure 6.2 are used in astronomical instruments. To minimise FRD (Section 4.2.5) it is necessary to feed fibres at close to their natural NA, but no faster. Therefore, feeding bare fibre at the native telescope focal plane is only efficient if the speed of the telescope focus approximately matches that of the fibre. Fortuitously, this is often the case at the prime focus of 4 m class telescopes, which were originally designed for photography. For example, the prime focus

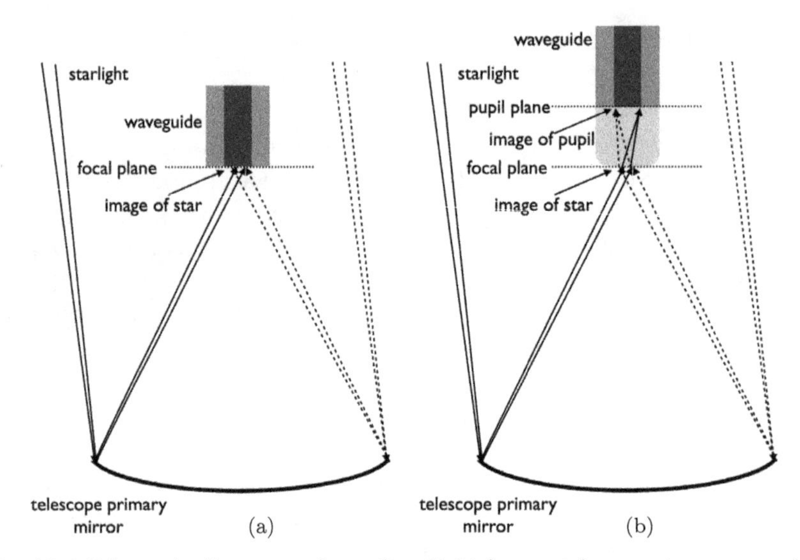

Fig. 6.2. Schematic diagrams of coupling light from a telescope to a waveguide via (a) the image plane or (b) the pupil plane.

speeds of the AAT, the Mayall Telescope and the Hale (200″) Telescope are $f/3.3, f/2.7.$ and $f/3.3$, respectively, which are well matched to fibres with $NA \approx 0.2 \approx f/2.5$. For situations when this is not the case, then either fore-optics must be used to magnify the beam, or microlenses on each fibre as in Figure 6.2(b) must be used.

Using microlenses has several advantages over bare fibres: (i) the input speed can be exactly matched to the fibre speed; (ii) microlenses can be hexagonal (or square) and therefore can be tessellated in an array to tile the focal plane without gaps. This allows spatially resolved spectroscopy of a contiguous region, e.g. a large nearby galaxy. A bare fibre array would have a lower filling factor (see Section 3.3.4 and Example 3.4); (iii) the microlens provides a pupil image on the front of the fibre, and the size of this pupil image can be matched to the size of the fibre diameter, providing efficient coupling (but note it is *not* possible to choose *any* combination of input focal ratio and pupil image diameter; the two are related through conservation of étendue; usually the input focal ratio will be chosen to minimise FRD, and then the fibre diameter will be matched to the resulting pupil diameter); and (iv) the pupil image will be approximately a top-hat profile (with a central hole), which matches the approximately top-hat profile of a multimode fibre.

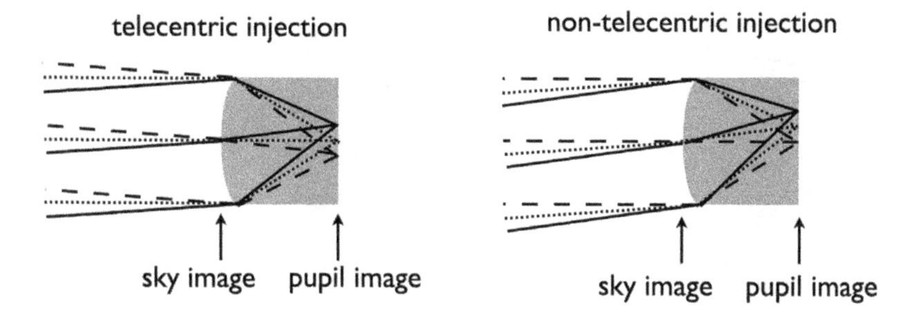

Fig. 6.3. Schematic diagram of a telecentric injection to a microlens and a non-telecentric injection, illustrating the shift of the pupil image.

A further consideration for efficient coupling is the need for telecentric injection into the fibres, i.e. the beam of light injected into the fibre must be aligned with the axis of the fibre. If this is not the case then there will be an increase in the maximum angle entering the fibre. By definition, the maximum angle of the beam entering the fibre will be increased by the non-telecentric angle. This increase in angle is known as geometric focal ratio degradation, since it is equivalent to feeding the fibre with a faster beam than the optics actually produce, and hence the output beam will be correspondingly faster. If using a microlens, there will be a shift of the pupil image formed on the front face of the fibre in addition to the increased angles entering the fibre; see Figure 6.3.

Geometric FRD can be avoided by ensuring that there is no non-telecentricity across the entire field, although such correction will usually require more lenses prior to the fibres. Another system to correct for geometric FRD when feeding fibres in the pupil plane uses double microlens as sketched in Figure 6.4. In this case, the sky image is formed on the back face of the first microlens. The second microlens serves to form a pupil image on the face of the fibre. This pupil image can be made telecentric since the distance between the sky image and the face of the second lenslet can now be adjusted such that it is equal to the focal length of the second lens.

6.2.2 *Coupling efficiency*

Having examined general schemes for coupling a waveguide to a telescope, we will now calculate the efficiency with which this can be achieved. The coupling efficiency is given by Equation 6.1 where E_0 is the electric field at the input face of the fibre. Therefore, E_0 will be different in

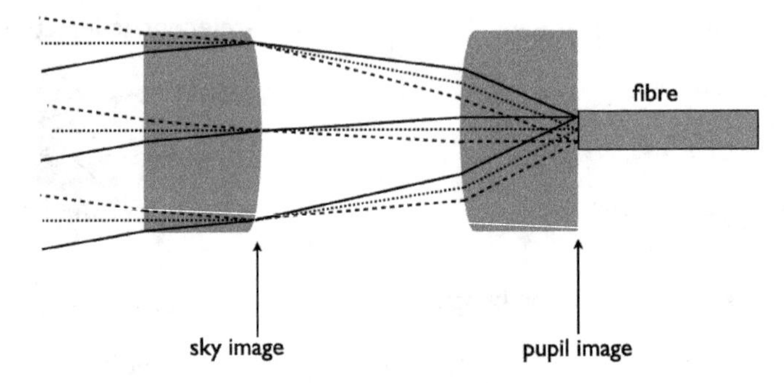

Fig. 6.4. A telecentric feed of a fibre with a double microlens.

the case of diffraction limited or seeing limited observations, and will be different in the image and the pupil planes.

The efficiency of this coupling is perhaps the major challenge of developing astrophotonic instruments, and as such deserves special consideration. This challenge originates in the competing requirements between single-mode photonic technologies which one wishes to adapt for astronomy, and multimode waveguides for efficient coupling of the light from the telescope. Let us see whence this conflict arises.

The étendue of a waveguide is given by

$$\begin{aligned} A\Omega &= \pi a^2 \times 2\pi(1-\cos\theta), \\ &= 2\pi^2 a^2 (1-\sqrt{1-NA^2}), \\ &\approx \pi^2 a^2 NA^2, \end{aligned} \tag{6.2}$$

where θ is the half-angle of the cone of light incident onto the face of the waveguide, and in the last line we have used $\sqrt{1-NA^2} \approx 1 - \frac{NA^2}{2}$. Therefore, the normalised frequency is

$$\begin{aligned} V &= 2\pi \frac{a}{\lambda} NA, \\ &\approx \frac{2}{\lambda}\sqrt{A\Omega}, \end{aligned} \tag{6.3}$$

and the corresponding number of modes is

$$M \approx \frac{V^2}{4} = \frac{A\Omega}{\lambda^2}. \tag{6.4}$$

Now, if étendue is conserved, then the $A\Omega$ in Equation 6.4 must be equal to that at the pupil of the telescope. If the telescope is being used in natural seeing conditions, with the fibre sampling a solid angle with a half-cone angle given by the HWHM of the seeing, i.e. $\Gamma/2$, from Equation 2.1, then the étendue is given by

$$A\Omega_{\text{tel}} = 2\pi^2 \left(\frac{D_{\text{tel}}}{2}\right)^2 \left(1 - \cos\frac{\Gamma}{2}\right)$$

$$A\Omega_{\text{tel}} \approx \left(\frac{\pi D_{\text{tel}}\Gamma}{4}\right)^2. \tag{6.5}$$

Example 6.1: How many modes would be required for a step-index fibre matched to 0.5 arcsec seeing on an 8 m telescope, at wavelengths of 1.5 μm and 0.5 μm?

From Equation 6.5 we find that $A\Omega_{\text{tel}} = 2.32 \times 10^{-10}$ m^2 sr. Putting this into Equation 6.4 gives, $M_{1.5} \approx 103$, and $M_{0.5} \approx 928$.

Example 6.2: What are the necessary core radii for the fibres of Example 6.1 if $NA = 0.2$?

Recalling Equations 5.9 and 5.10, we find

$$a = \frac{V\lambda}{2\pi NA}$$

$$= \frac{\sqrt{M}\lambda}{\pi NA},$$

and therefore, we find that $a_{1.5} = 24$ μm and $a_{0.5} = 73$ μm.

These results are independent of the presence or absence of any fore-optics to magnify the beam, or any microlenses to project a pupil image; the results are the consequence of the conservation of étendue and must always hold so far as the optics conserve étendue.

Note that the single-mode requirement of $V < 2.405$ is approximately correct for *any* waveguide, including endlessly single-mode PCF. Therefore the requirement to sample the seeing disc on the sky (Section 2.1), in order to collect sufficient photons from a point source, and to feed the fibre at less than the maximum *NA*, in order to avoid loss in the injection, is incommensurate with the requirement for single-mode injection into photonic devices. This problem presents challenges for astrophotonics, but it is not intractable, as will be discussed in the following section. Before looking at solutions to this problem, let us examine the coupling efficiency between a telescope and a fibre in more detail.

6.2.2.1 *Diffraction limited coupling in the single and few mode regimes*

We will begin with a calculation of the coupling efficiency in the diffraction limit, in the single-mode and few mode regimes. In practice it is extremely difficult for a ground-based telescope to achieve truly diffraction limited performance due to the seeing (Section 2.1), especially at visible or near-infrared wavelengths. Nevertheless, this calculation is instructive, both heuristically, and because it represents the ultimate coupling efficiency of any system.

Shaklan and Roddier (1988) examined coupling into single-mode fibres in natural seeing and the diffraction limit, in both the image and pupil planes. Bland–Hawthorn and Horton (2006) extended this analysis to look at coupling into few-mode fibres in diffraction limited conditions in the image plane. Here, we closely follow these papers, and Ellis *et al.* (2021) to extend the analysis to include few mode fibres in both diffraction limited and seeing limited conditions, and both the image and pupil planes.

Image plane. In general we may describe the electric field in the pupil plane by

$$E_{\mathrm{pupil}}(\mathrm{r}) = E_{\mathrm{S}} P(\mathrm{r}) \Psi(\mathrm{r}), \tag{6.6}$$

where E_{S} is the electric field amplitude of the source, $P(\mathrm{r})$ is the pupil transmission function, and $\Psi(\mathrm{r})$ is a turbulent phase screen, which takes into account distortions in the electric field due to atmospheric turbulence, and optical aberrations, etc.

In the image plane the electric field is given by the Fourier transform of E_{pupil}, i.e.

$$E_{\mathrm{image}}(\mathrm{r}) = \widehat{E}_{\mathrm{pupil}}\left(\frac{\mathrm{r}}{\lambda F D_{\mathrm{tel}}}\right), \tag{6.7}$$

where F is the focal ratio at the image plane (Equation 2.5), D_{tel} is the diameter of the telescope, and λ is the wavelength.

The pupil function of a circular telescope aperture of diameter D with a central obstruction of diameter αD (neglecting the effect of the 'spider' vanes, which hold the secondary mirror in place) is simply

$$P(\mathrm{r}) = \begin{cases} 1, & \frac{\alpha D}{2} \leq |\mathrm{r}| \leq \frac{D}{2}, \\ 0, & \text{otherwise.} \end{cases} \tag{6.8}$$

In the diffraction limited case $\Psi(\mathrm{r}) = 1$, and therefore Equation 6.7 becomes

$$E_{\mathrm{image}} = E_{\mathrm{S}} \left(\frac{2J_1(\mathrm{s})}{\mathrm{s}} - \alpha^2 \frac{2J_1(\alpha \mathrm{s})}{\alpha \mathrm{s}} \right), \tag{6.9}$$

where

$$\mathrm{s} = \frac{\pi \mathrm{r}}{\lambda F}. \tag{6.10}$$

The coupling efficiency is then found by Equation 6.1, with E_0 given by E_{image} above. Bland–Hawthorn and Horton (2006) calculated the coupling efficiency in the case of diffraction limited image plane for various combinations of fibre numerical aperture, fibre diameter and focal ratio injection. We have repeated some of their calculations, as shown in Figure 6.5. Note that the maximum coupling efficiency for a single-mode fibre is approximately 80% (in agreement with the analysis of Shaklan and Roddier 1988); this is the best possible coupling efficiency for a standard step-index SMF under ideal conditions. Higher coupling efficiency is possible if the fibre is designed to have the same electric field distribution as the input light. For example, Gris-Sánchez *et al.* (2016) designed and made a fibre to accept the Airy pattern from a diffraction limited circular aperture, with a coupling efficiency of 93.7%.

Pupil plane. At first sight, coupling in the pupil plane seems slightly more complicated, since there is an extra optical element involved (usually a microlens bonded to the fibre end-face, as in Figure 6.2) in order

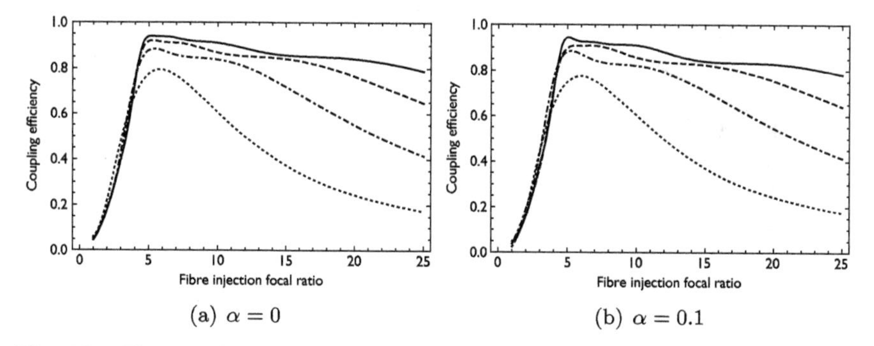

Fig. 6.5. The coupling efficiency in the image plane at a wavelength of 1.55 μm into an $NA = 0.1$ fibre as a function of input focal ratio for fibres of 10, 25, 40 and 50 μm diameter (dotted, dot-dashed, dashed and continuous respectively), following Bland–Hawthorn and Horton (2006). The left panel is for a telescope with no central obstruction, and the right panel for a telescope with a central obstruction of $\alpha = 0.1$ by diameter.

to re-image the pupil onto the fibre. Indeed, to be rigorous one should take the Fourier transform of the microlens pupil multiplied by the incident electric field to yield the pupil image on the front face of the fibre (see, e.g. Corbett 2009). Here however, we make the simplifying approximation that the pupil image will just be a magnified image of the pupil itself, and do not take into account any diffraction effects from the reimaging optics. Therefore, in the diffraction limit the electric field at the pupil plane is simply

$$E(\mathrm{r}) = E_{\mathrm{S}} P\left(\frac{\mathrm{r}}{\lambda F D_{\mathrm{tel}}}\right). \tag{6.11}$$

Figure 6.6 shows the coupling efficiency in the pupil plane as a function of fibre diameter for a fibre with $NA = 0.1$ at a wavelength of 1.55 μm when the pupil image from an 8 m telescope (with no central obstruction) is matched to the fibre NA, i.e. the pupil image is injected at the ideal NA, but therefore the pupil diameter is *not* matched to the fibre diameter. In this case, the pupil image size is 18.3 μm, and indeed we see a peak in the coupling efficiency around this fibre diameter, and smaller coupling efficiency for smaller fibres when a lot of the light is lost. For larger diameters the coupling remains good since the light can still be captured by the fibre, but in practice would lead to an increase in étendue, as the emerging

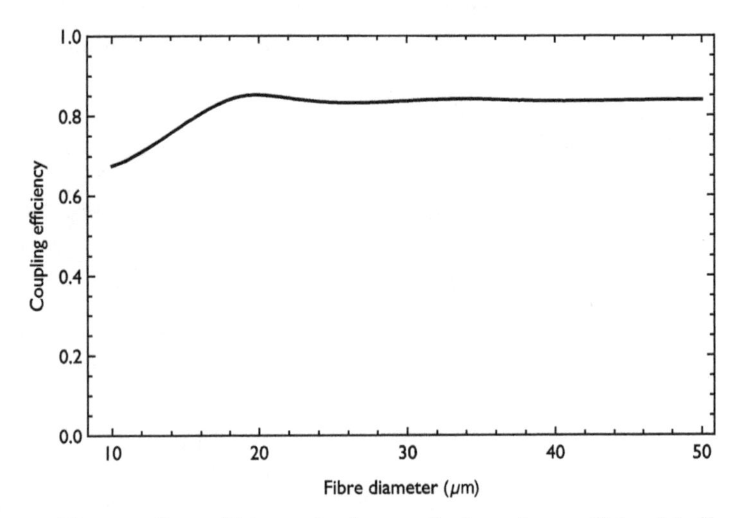

Fig. 6.6. The coupling efficiency in the pupil plane for an $NA = 0.1$ fibre as a function of fibre diameter at a wavelength of 1.55 μm for an 8 m telescope, injecting at a focal ratio equal to the fibre NA.

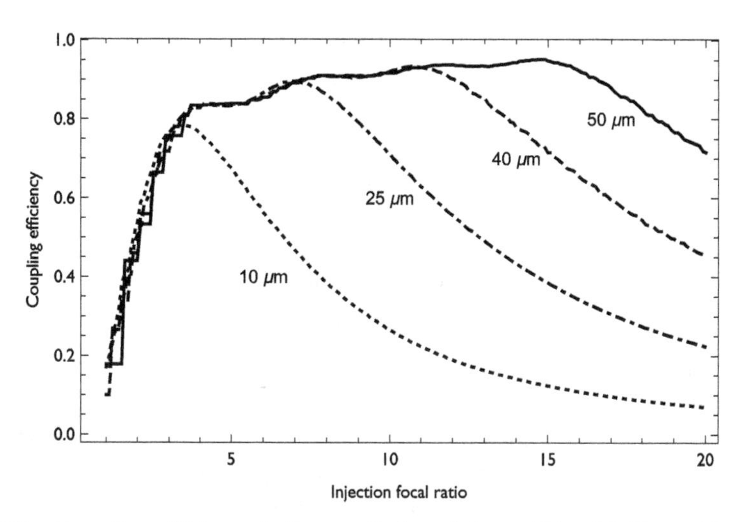

Fig. 6.7. Coupling efficiency in the pupil plane into an $NA = 0.1$ fibre with a core diameter of 10, 25, 40, and 55 μm, as a function of input focal ratio from an 8 m diameter telescope with no central obstruction.

spot size would be larger than the input due to cross-talk in the modes, equivalent to the effects of FRD (Section 4.2.5).

Figure 6.7 shows the results for the same fibre and telescope, but as a function of input injection focal ratio, for fibre diameters of 10, 25, 40, and 50 μm.

6.2.2.2 *Coupling in seeing limited conditions in the single-mode case*

We now turn our attention to coupling in seeing limited cases, beginning with the single-mode case. Atmospheric turbulence will distort the incoming wavefront as described in Section 2.1, making coupling between the telescope and fibre less efficient.

We can calculate this efficiency by adding a phase term to the electric field distribution across the pupil, and calculating the overlap integral with the fibre modes as for the diffraction limited case. To compute the phase screen across the pupil for particular seeing conditions we begin by assuming that the turbulence can be approximated by a **Kolmogorov power spectrum**, when the phase is given by

$$\phi(\kappa) = 0.023 r_0^{-\frac{5}{3}} \kappa^{-\frac{11}{3}}, \tag{6.12}$$

where r_0 is Fried's parameter (Equation 2.1) and κ is the spatial frequency. The variance of the difference in phase across the pupil is called the **structure function**, and is given by

$$D(r) = 6.88\left(\frac{r}{r_0}\right)^{\frac{5}{3}}. \tag{6.13}$$

The distance between two points in the pupil plane, r, is related to the spatial frequency, the wavelength, λ and the focal ratio, f, by

$$r = \lambda f \kappa. \tag{6.14}$$

See, e.g. Hardy (1998) for more details.

For example, Figure 6.8 shows example wavefronts (panel a) across an 8 m pupil for seeing of 0.3, 0.5, and 0.9 arcsec, and the resulting **speckle image** (panel b), as well as the average image from 100 realisations (panel c), and the best-fitting Moffat function (panel d). The speckle image is the instantaneous image, which is formed from the Fourier transform of the turbulent wavefront across the pupil, as given in Equation 6.7. The incoherent wavefront results in an image which is 'speckled' due to the random constructive and destructive interference.

Pupil plane: For coupling in the pupil plane we take the electric field across the pupil as calculated according to the above algorithm, and scale it to the appropriate pupil diameter for a given injection focal ratio into the fibre. The scaling is done by conserving étendue (Section 2.4), given the size of the seeing on the sky, the diameter of the telescope, and the desired focal ratio at the fibre input.

Example 6.3: Calculate the pupil diameter for a fibre with $NA = 0.2$ being fed at this NA from a microlens with refractive index of 1.5, if the fibre is to subtend 0.5 arcsec on the sky, and is fed from an 8 m telescope.
Using the linear étendue, we have

$$\left(\frac{d_{\text{tel}}}{2}\right)\sin\theta_{\text{sky}} = \left(\frac{d_{\text{fibre}}}{2}\right) n \sin\theta_{\text{pupil}}$$

$$d_{\text{tel}}\sin\theta_{\text{sky}} = d_{\text{pupil}} NA$$

$$8\sin\left(\frac{0.25}{3600}\frac{\pi}{180}\right) = d_{\text{pupil}} NA \tag{6.15}$$

$$\implies d_{\text{pupil}} = 48.5\ \mu\text{m}. \tag{6.16}$$

Therefore, for good coupling efficiency, and allowing for some misalignment, and fibre with a core diameter slightly larger than 48.5 μm should be chosen.

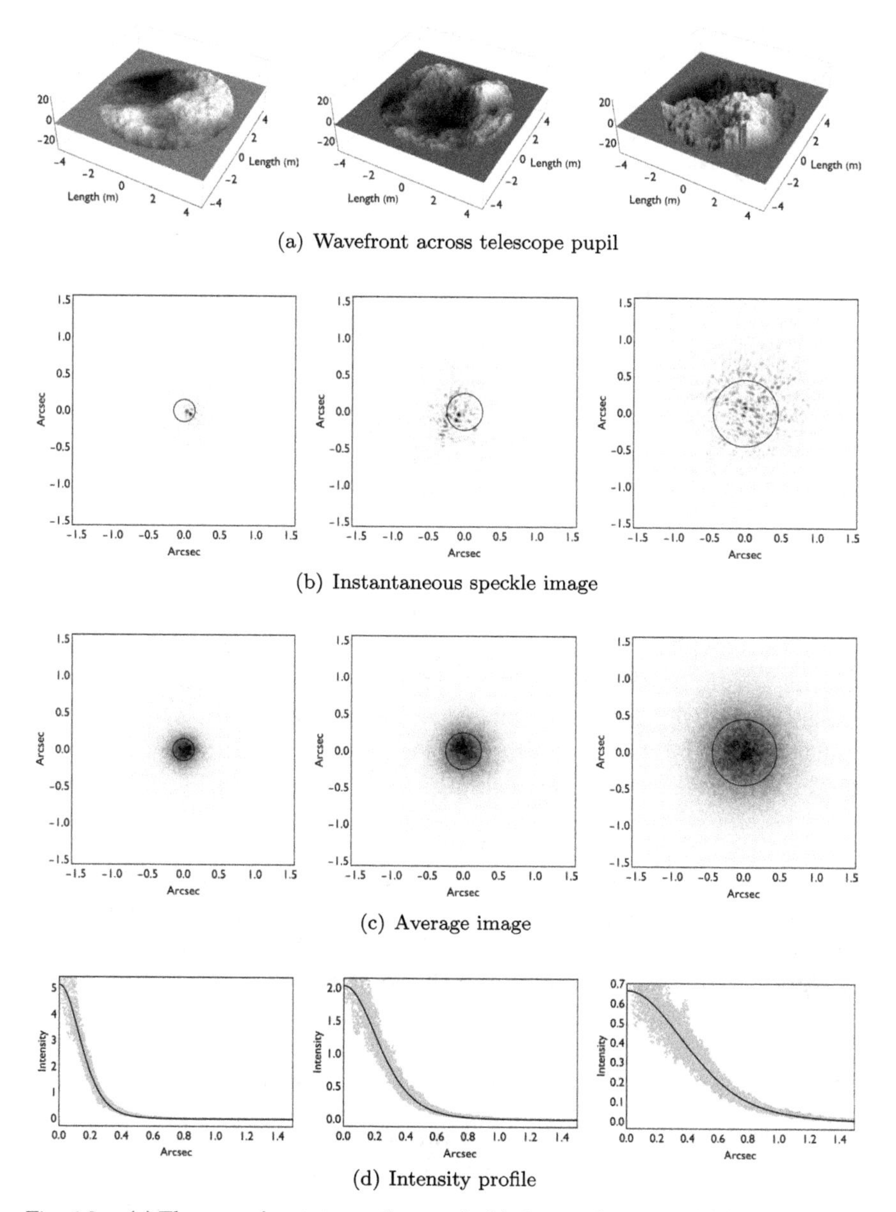

Fig. 6.8. (a) The wavefront across the pupil; (b) the resulting speckle image; (c) the average image from 100 speckle images; and (d) the best fitting Moffat function, for seeing on 0.3 arcsec (left), 0.5 arcsec (middle), and 0.9 arcsec (right) across an 8 m pupil.

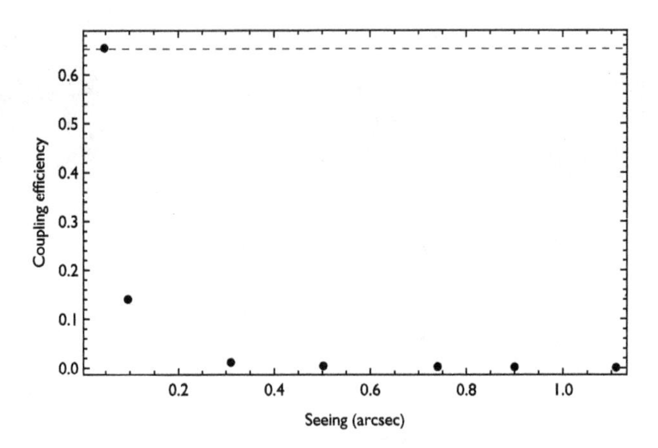

Fig. 6.9. The coupling efficiency for pupil imaging from an 8 m telescope into a single-mode fibre for seeing of 0.001 (i.e. diffraction limited), 0.3, 0.5, 0.7, 0.9, and 1.1 arcsec at a wavelength of 1.44 μm. The dashed line shows the analytic diffraction limited coupling.

Figure 6.9 shows the efficiency of coupling from an 8 m telescope into a standard SMF-28 fibre with a core diameter of 8.2 μm and an *NA* of 0.14 for seeing of 0.001 (i.e. diffraction limited), 0.3, 0.5, 0.7, 0.9, and 1.1 arcsec at a wavelength of 1.55 μm. This is calculated for the average of 100 realisations of the seeing, as shown in Figure 6.8(a). Even for exceptionally good seeing of 0.3 arcsec the coupling efficiency is only $\approx$1%, and at more typical seeing of >0.5 arcsec the coupling efficiency drops to nearly zero.

Image plane. We follow the same procedure for coupling in the image plane, this time averaging the coupling from 100 realisations of the instantaneous speckle image as shown in Figure 6.8(b). The results are shown in Figure 6.10. Again, for any realistic seeing the coupling efficiency is nearly zero.

6.2.2.3 *Coupling in seeing limited conditions in the multimode case*

For coupling into multimode fibre the same procedure can be followed as for the single-mode case, and taking the sum of the overlap integral for each mode as in Equation 6.1. However, such computations become very lengthy for a large number of modes, and a good approximation can be calculated much more simply by simply taking the ratio of étendue of the fibre injection to the étendue at the telescope entrance pupil, as in the following example.

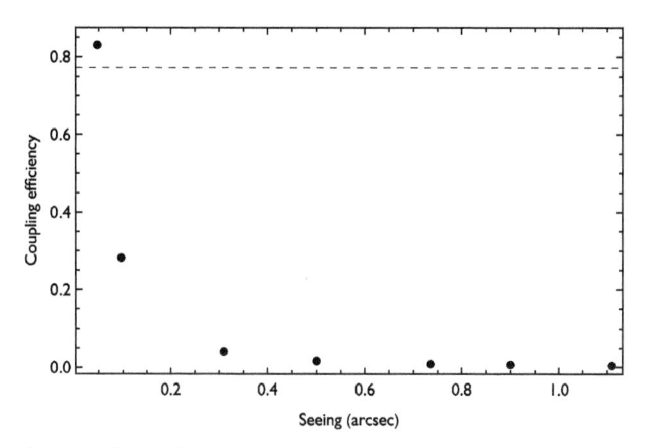

Fig. 6.10. The coupling efficiency in the image plane from an 8 m telescope into a single-mode fibre for seeing of 0.001 (i.e. diffraction limited), 0.3, 0.5, 0.7, 0.9, and 1.1 arcsec. The dashed line shows the analytic diffraction limited coupling.

Example 6.4: From the ratio of étendue, calculate the coupling efficiency from an 8 m telescope and a seeing of 0.6 arcsec into a fibre with a core diameter of 50 μm and an NA of 0.22.
The area of the telescope entrance pupil, neglecting the central obstruction, is simply

$$A_{\text{tel}} = \pi \left(\frac{d_{\text{tel}}}{2} \right)^2 = 50.27 \text{ m}^2. \tag{6.17}$$

The solid angle subtended by the seeing is

$$\Omega_{\text{tel}} = 2\pi \left(1 - \cos \frac{\Gamma}{2} \right) = 0.28 \text{ arcsec}^2. \tag{6.18}$$

Meanwhile, at the fibre the area is simply,

$$A_{\text{fibre}} = \pi \left(\frac{d_{\text{fibre}}}{2} \right)^2 = 1.964 \times 10^{-9} \text{ m}^2, \tag{6.19}$$

and the solid angle when the fibre is injected at its NA is

$$\Omega_{\text{fibre}} = 2\pi \left(1 - \cos \left(\sin^{-1} NA \right) \right) = 6.55 \times 10^9 \text{ arcsec}^2. \tag{6.20}$$

Therefore, the coupling efficiency is

$$\eta = \frac{A_{\text{fibre}}\Omega_{\text{fibre}}}{A_{\text{tel}}\Omega_{\text{tel}}},$$
$$= 0.90. \tag{6.21}$$

N.B. This efficiency does not account for aperture losses; the full width at half-maximum (FWHM) of a Moffat function only encloses ≈43% of the total light from a point source (depending on the β parameter), and so the actual coupling efficiency should be modified by this factor if matching the fibre to the seeing FWHM, giving an efficiency of 0.39.

Although, the efficiency in Example 6.4 is only 0.39, largely due to aperture losses, when designing a fibre system the optimal fibre size, or the optimal angular size of the fibre on the sky, should be found by optimising the signal-to-noise (Equation 3.1); although increasing the sampling on the sky will minimise aperture losses it will also increase the sky background.

Figure 6.11 shows the coupling efficiency for an 8 m telescope as a function of fibre diameter for seeing of 0.3 and 0.5 arcsec at a wavelength of 1.5 μm, for fibres with $NA = 0.1$ and 0.2, and an input injection focal ratio matched to the fibre NA. Note well that the approximate method based on the ratio of étendue agrees well with the numerical method based on the calculation of the wavefronts. Since the angular size of the fibre on the sky is defined to be equal to the FWHM of a Moffat profile, the maximum coupling efficiency is given by the ratio of the light contained within the FWHM, which is ≈0.38, as in Example 6.4.

Figure 6.12 shows the same calculations for the ratio of étendue, taking into account aperture losses, but for coupling in the image plane, for a seeing of 0.3, 0.5, and 0.7 arcsec.

To reiterate, in order to achieve high coupling efficiency in seeing limited conditions it is necessary to use fibres with a large étendue, which in practice means a larger core diameter, leading to a high number of modes. In practice, the fibre diameter and NA will often be selected to ensure high coupling efficiency by matching the étendue given by the desired sampling on the sky, and the telescope collecting area.

Figure 6.13 shows the number of modes in a step-index fibre matched to the telescope diameter and seeing, for wavelengths of 0.5 and 1.5 μm. Hundreds or thousands of modes are common in astronomical use.

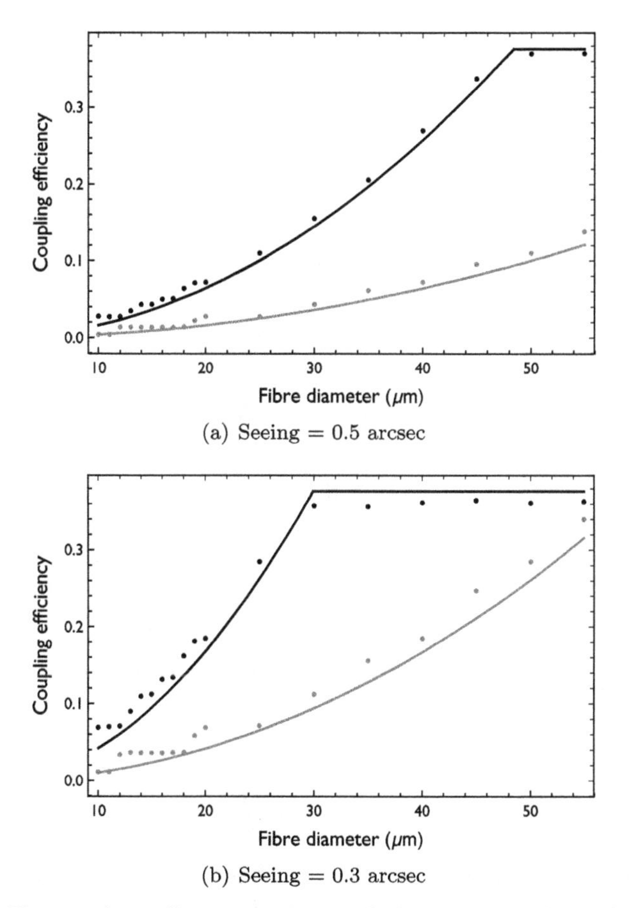

(a) Seeing = 0.5 arcsec

(b) Seeing = 0.3 arcsec

Fig. 6.11. The coupling efficiency in the pupil plane in seeing limited conditions, for fibres with $NA = 0.1$ (grey points) and $NA = 0.2$ (black points) for an injection focal ratio matched to the fibre NA, as a function of fibre diameter. Overlaid are curves estimating the coupling efficiency from the ratio of the fibre étendue to the input étendue.

6.3 Coupling to Single-Mode Photonics

We have shown that efficient coupling of a fibre or waveguide in seeing limited conditions necessarily requires the use of a multimode waveguide, and we have examined in detail the achievable coupling efficiencies. However, many astrophotonic devices require the use of single-mode

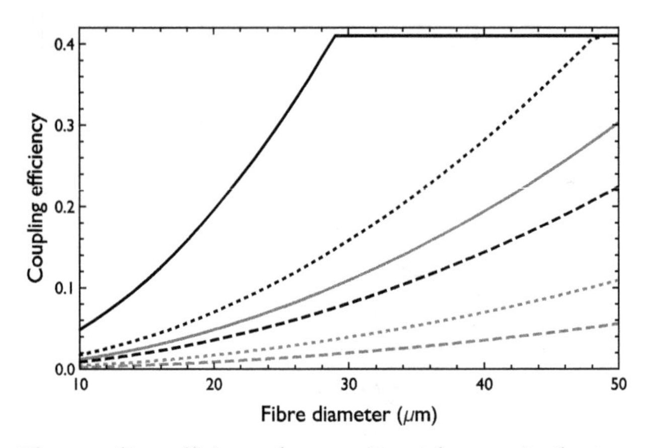

Fig. 6.12. The coupling efficiency from an 8 m telescope in the image plane in seeing limited conditions, for fibres with $NA = 0.1$ (grey lines) and $NA = 0.2$ (black lines) for an injection focal ratio matched to the fibre NA, as a function of fibre diameter, at a seeing of 0.3 (continuous), 0.5 (dotted), and 0.7 (dashed) arcsec.

waveguides to function correctly. We will now examine how high coupling efficiency with single-mode waveguides may be achieved.

6.3.1 *Adaptive optics*

In Section 6.2.2.1, we saw that high coupling efficiency into a single-mode fibre could be achieved if working at the diffraction limit. One way to achieve this is in seeing limited conditions is to correct the effects of atmospheric turbulence with adaptive optics (Section 2.5) to reproduce images which are close to diffraction limited.

If the correction is very good, so called **extreme adaptive optics** or **ExAO**, then the coupling efficiency into a single-mode fibre can be quite high. Jovanovic *et al.* (2017) have demonstrated a coupling efficiency of >40% for greater than 84% of the time using the Subaru Coronagraphic Extreme AO (SCExAO) instrument on the Subaru telescope, and calculate that this could be pushed as high as 67% with improvements in the ExAO wavefront correction.

Adaptive optics is now a mature and very diverse field. The performance of AO will certainly continue to improve as advances in wavefront sensing, and laser guide stars progress. Investment into AO will also be very important for the success of the next generation of extremely large telescopes (ELTs), which seek to exploit the superior diffraction limited

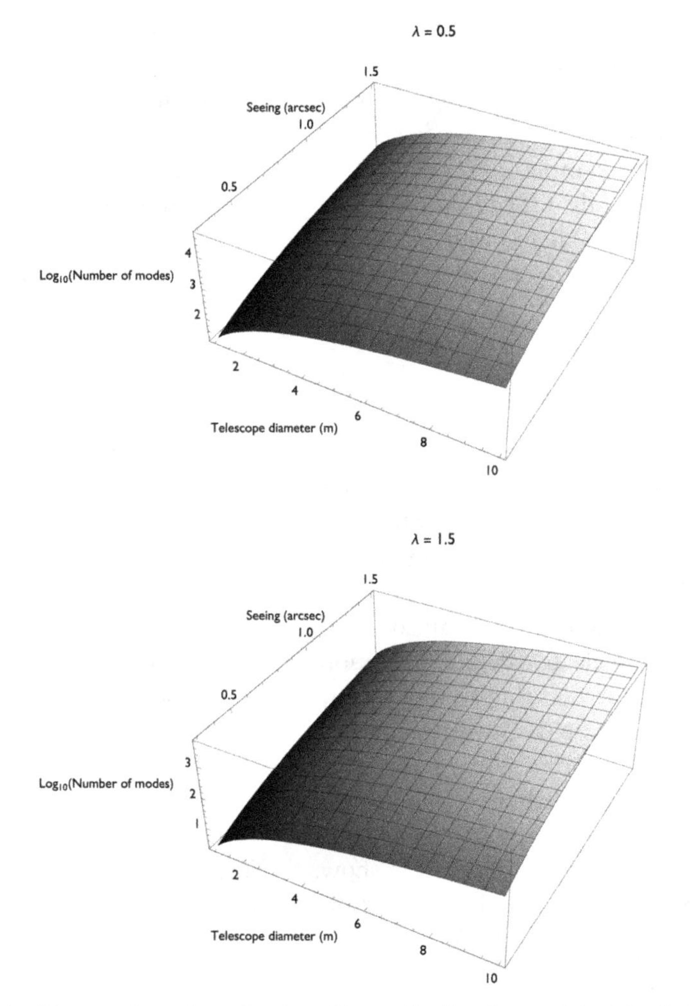

Fig. 6.13. The number of modes in a fibre with étendue matched to that of the telescope collecting area and seeing conditions.

performance of their very large mirrors with AO, especially for goals such as exoplanet detection and imaging.

The improvements and proliferation of AO systems on large telescopes will be very beneficial for astrophotonic instruments, making it much easier to inject into single-mode waveguides. However, only the extreme AO systems, such as the one used by Jovanovic *et al.* (2017) will achieve good enough correction to inject directly into SMF. Other branches of AO with different priorities, e.g. correcting over larger fields-of-view will be of less benefit. Furthermore, AO systems come with some compromises, though these are becoming fewer and less detrimental as the field progresses. For example, a bright 'natural guide star' is always necessary for the low-order tip-tilt correction, thereby reducing the fraction of the sky which is accessible to observation. AO instruments are also very complicated with many optical elements, which reduces the overall throughput of the system.

6.3.2 *The photonic lantern*

Another solution to the conflict between the requirements of single-mode photonics and multimode waveguides for collecting adequate light at the telescope is found in the **photonic lantern**. This device efficiently converts a multimode input to an array of single-mode fibres, or vice versa (Leon-Saval *et al.* 2005).

Photonic lanterns operate on the principle of tapering waveguides such that the electric field distribution becomes less confined to the core; as the core narrows throughout the taper more of the light is contained in the evanescent field. Considering a circular step-index fibre, recall that the electric field distribution is given by Equations 5.1–5.9. Figure 6.14 shows how the field distribution evolves as the core becomes narrower for an example fibre. As the core narrows more of the field is evanescent, extending into the cladding. Figure 6.15 shows the fraction of the intensity which is in the evanescent field as a function of core size for a silica fibre with a core index difference of 0.01.

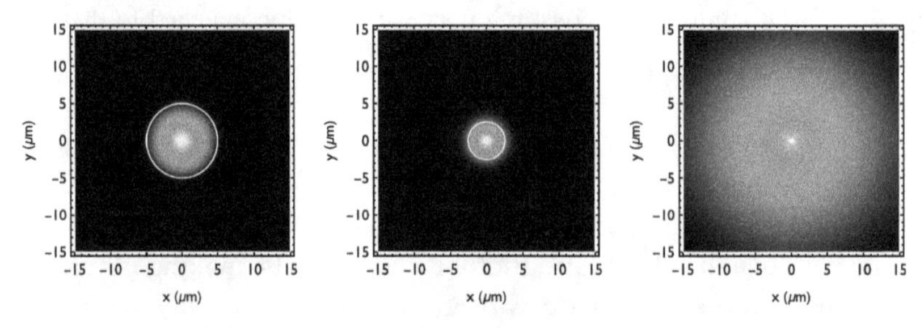

Fig. 6.14. Plots of the intensity of fundamental mode of a step-index fibre, for different core sizes (depicted by the white circles). As the core narrows more of the field is evanescent, extending into the cladding.

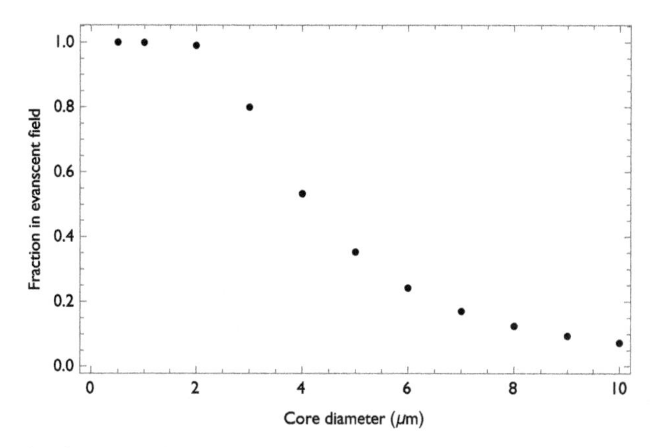

Fig. 6.15. The fraction of the intensity which is in the evanescent field as a function of core size for a silica fibre with a core index difference of 0.01.

Now consider the arrangement of waveguides shown in Figure 6.16. At the top of the diagram an array of single-mode fibres is housed within a low-index jacket, and each fibre operates as a standard single-mode fibre. The fibres and jacket are made to undergo an adiabatic taper, by which we mean that the taper is slow enough that the transition is lossless and reversible. As the cores of the fibres narrow, more of the light is squeezed out of the core into the evanescent field as for Figures 6.14 and 6.15. Eventually, all the light is contained within the cladding of the single-mode fibres, which now acts as the core of multimode fibre, with the low index jacket forming the new cladding. The multimode end of the fibre carries the M modes from the M single-mode cores. N.B. this process is necessarily reversible, since Maxwell's equations are symmetric with respect to time, and so light can also be injected into the multimode end, whereafter it will begin to be guided more strongly by the cores of the single-mode fibres, as they grow in size along the taper, until all the light is contained in the single-mode fibres. The device must be designed such that the multimode end matches the number of single-modes.

This device is known as a photonic lantern.[1] This device is a cornerstone of astrophotonics, since it allows single-mode photonic devices to be successfully and efficiently interfaced with multimode fibres, and therefore with a telescope.

[1]The name originated from the similarity between a diagram of an early multimode–single-mode–multimode device and a Chinese paper lantern.

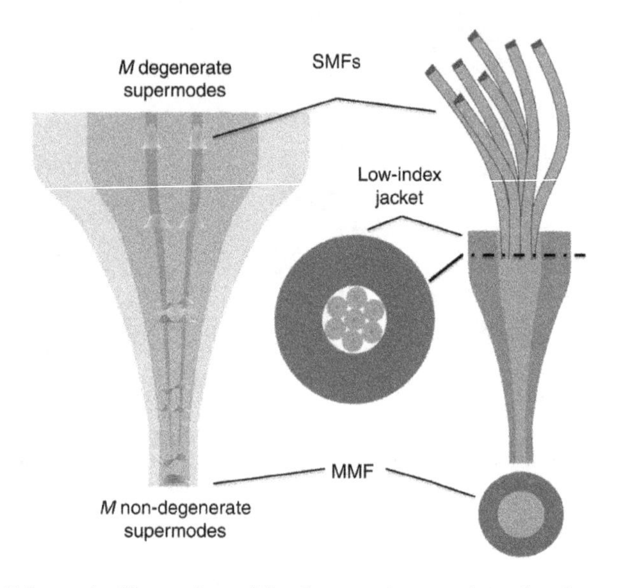

Fig. 6.16. Schematic illustration of the form and operation of a photonic lantern.

Standard photonic lanterns made from arrays of single-mode fibres have been written with up to 88 modes (Birks *et al.* 2015; this has been bettered using multi-core fibre, see Section 6.3.2.1). Efficiencies >90% have been achieved for few-mode devices (Birks *et al.* 2015). The efficiency depends on a close matching of the number of modes in the multimode input to the number of single-mode cores, but the throughput is not a trivial function of injection focal ratio, and should be properly measured and characterised when designing an astronomical instrument (Horton *et al.* 2014).

Without photonic lanterns single-mode astrophotonics would have to rely on adaptive optics correction to couple successfully to a telescope beam, or to suffer unacceptably large losses. For astrophotonics the photonic lantern has some advantages over an AO injection, the most obvious being that it does not require a complicated and expensive AO system. Moreover, a photonic lantern does not require the use of a guide star and therefore has full sky-coverage, unlike AO. These advantages are limited to spectroscopy, since a photonic lantern cannot recover the diffraction limited image, which is of course possible with AO. Also, since the number of modes in a fibre varies as $1/\lambda^2$, the lantern has a wavelength dependent efficiency that needs to be taken into account.

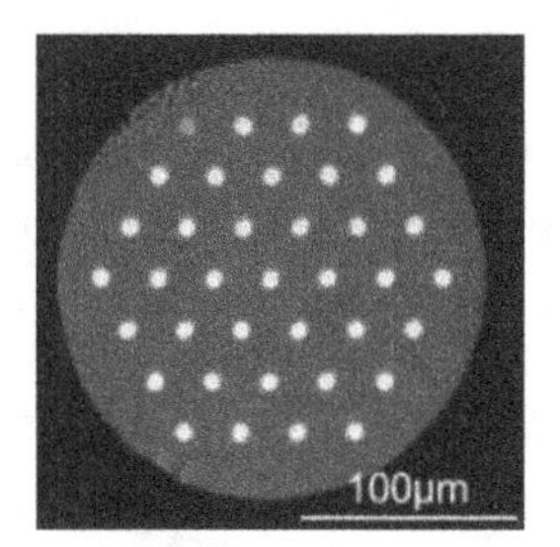

(a) Multicore fibre

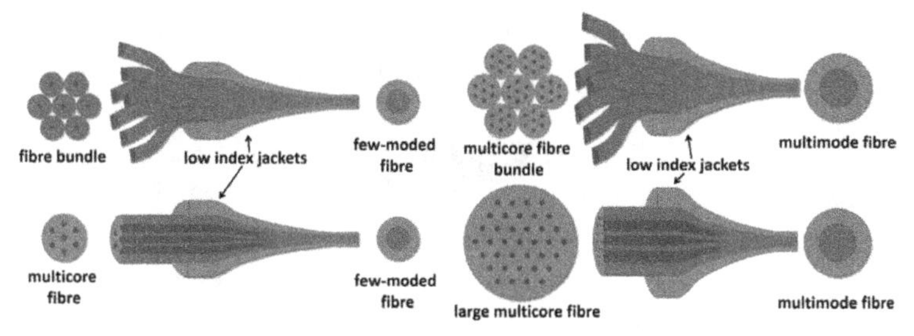

(b) Multicore fibre photonic lantern

Fig. 6.17. (a) A photograph of the end-face of a 37-core multicore fibre; (b) a comparison of different schemes for producing photonic lanterns: a bundle of individual SMFs (top left), an individual multicore fibre (bottom left), a bundle of multicore fibres (top right), and a large individual multicore fibre (bottom right). All figures are from Leon-Saval *et al.* (2017).

6.3.2.1 *Multicore fibre photonic lanterns*

A similar scheme may be realised using **multicore fibre (MCF)** instead of individual single-mode fibres. A multicore fibre, as the name suggests, contains not a single core surrounded by the cladding, but many cores, e.g. Figure 6.17. A multicore fibre may be inserted into a low-index jacket and tapered to form a photonic lantern in the same way as described above. Devices with up to 121 cores have been written (Birks *et al.* 2015). An advantage of multicore fibres is that they are less cumbersome to work with than individual fibres, since they can be handled as a single fibre would be.

Leon-Saval *et al.* (2017) have suggested a hybrid-scheme based on a bundle of multi-core fibres, see Figure 6.17(b). This 'divide-and-conquer' scheme allows the number of modes incorporated to be increased, with devices having up to 259 modes already demonstrated.

6.3.2.2 *Direct-write photonic lanterns*

Photonic lanterns have also been written using direct-write waveguides (Spaleniak 2014; Spaleniak *et al.* 2012, 2013; Thomson *et al.* 2011, 2012, 2009). In this case, the transition from multimode to single-mode behaviour is not achieved through tapering the cores, but rather they are just brought closer together until the modes are strongly overlapping, and the individual cores act as a single multimode core carrying M supermodes (see Section 6.6), where M is the number of single-mode cores.

This method has some advantages. First, the layout can be arbitrary, for example going from a multimode input to an output slit, where the slit can be a 'true slit', since the cores do not have to be separated by the width of any fibre cladding. Second, they can be written simultaneously with other devices, such as Bragg gratings (Spaleniak *et al.* 2014) or AWGs (Douglass *et al.* 2018). Finally, they are small and robust, since they do not contain any loose fibres. Unfortunately, the throughput is not yet as good as fibre-based photonic lanterns, and they have not been demonstrated for a large number of modes.

6.4 Coupling from Waveguides into Astronomical Instruments

Having examined in detail the problems and solutions to coupling light from a telescope into a waveguide, we now turn our attention to the reciprocal problem, i.e. that of coupling the light from the waveguides into an instrument.

When coupling from fibres to a conventional spectrograph it is normal to arrange the fibres into a pseudo-slit, i.e. a linear array of the fibres. Light can then be injected into the instrument as from a real slit. Note that the 'pseudo-slit' can be curved (in the direction of the optical axis) to aid with the optical design of the spectrograph. The fibre output can either be imaged directly, or else the pupil image can be converted to a sky-image with output microlenses (equivalent to the situation in Figure 6.2).

Note that when using a photonic lantern to enable single-mode behaviour in the fibres it is often required to convert back to a multimode output before forming a pseudo-slit. The reason for this is that the light injected into the spectrograph is then spread over fewer pixels, and therefore this minimises detector background noise.

6.4.1 *Coupling from a fibre into a spectrograph*

A very common use of fibres in astronomy is in multi-object spectroscopy (MOS; Section 3.3.3). We will use this as a heuristic example to introduce the issues related to coupling fibres into instruments, first looking at the conventional use of multimode fibre for MOS, then looking at single-mode spectrographs, following the treatment given in Robertson and Bland–Hawthorn (2012).

6.4.1.1 *Multimode fibre*

Consider the sketch of a fibre fed spectrograph shown in Figure 6.18, in which the output of a bare fibre is collimated onto a diffraction grating, the output of which is then imaged onto a detector.

The grating equation relates the grating period b, the angle of incidence, θ_{i}, the angle of diffraction, θ, to the wavelength, λ, and the diffraction order m by

$$b \sin \theta_{\mathrm{i}} + b \sin \theta = m\lambda. \tag{6.22}$$

Differentiating with respect to λ gives the angular dispersion

$$\frac{\Delta\theta}{\Delta\lambda} = \frac{m}{b \cos \theta}. \tag{6.23}$$

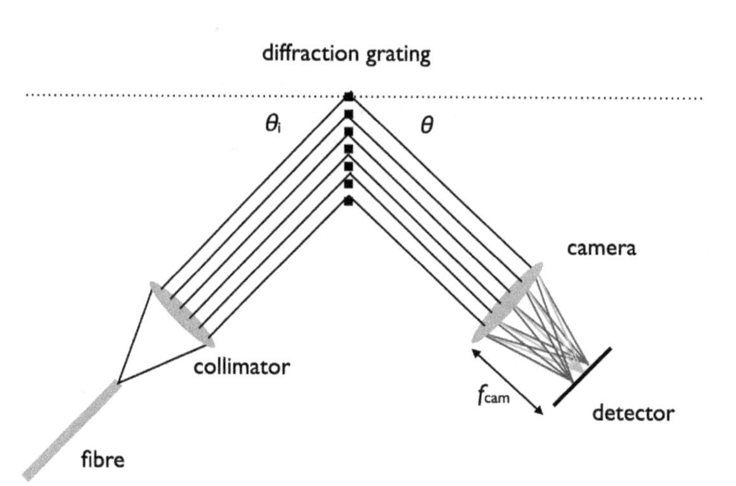

Fig. 6.18. Sketch of a fibre fed spectrograph. The output from a bare fibre is collimated, diffracted, then imaged onto a detector.

In the Littrow configuration the angle of diffraction at the centre wavelength is equal to the angle of incidence, i.e.

$$2b \sin \theta_{\rm i} = m\lambda_{\rm c}. \tag{6.24}$$

In this case, Equation 6.23 becomes

$$\Delta\theta = 2\frac{\Delta\lambda}{\lambda_{\rm c}} \tan\theta. \tag{6.25}$$

Therefore, the linear dispersion at the detector will be

$$\Delta x = 2f_{\rm cam}\frac{\Delta\lambda}{\lambda_{\rm c}} \tan\theta_{\rm i}, \tag{6.26}$$

where $f_{\rm cam}$ is the focal length of the camera.

Now, consider the output from a highly multimode fibre, and let us assume that diffraction effects from the fibre itself are negligible. In this case the illumination of the diffraction grating will be that of a circular aperture, and the amplitude of the point spread function (PSF) will be given by the Fourier transform of this (see Equation 6.9), which in the image plane of the detector is given by

$$E_{\rm psf}(s) = \frac{2J_1(s)}{s}, \tag{6.27}$$

$$s = \frac{\pi d r}{\lambda f_{\rm cam}}, \tag{6.28}$$

where d is the diameter of the beam on the diffraction grating.

The intensity at the image plane is given by the absolute square of this, i.e. an Airy function, as sketched in Figure 6.19(a). The Rayleigh criterion states that two Airy functions are considered to be just resolved when the maximum intensity of one overlaps with the first minimum in intensity of the other, as sketched in Figure 6.19(b).

Equating the Rayleigh criterion with the linear dispersion at the detector we find that the resolving power is

$$R_{\rm MMF} = \frac{\lambda_{\rm c}}{\Delta\lambda} = \frac{2}{1.22}\frac{d}{\lambda}\tan\theta_{\rm i}. \tag{6.29}$$

Note, for a rectangular slit the number of illuminated lines on the grating is

$$\begin{aligned} N &= \frac{d}{b} \\ &= \frac{2d\sin\theta_{\rm i}}{m\lambda_{\rm c}}, \end{aligned} \tag{6.30}$$

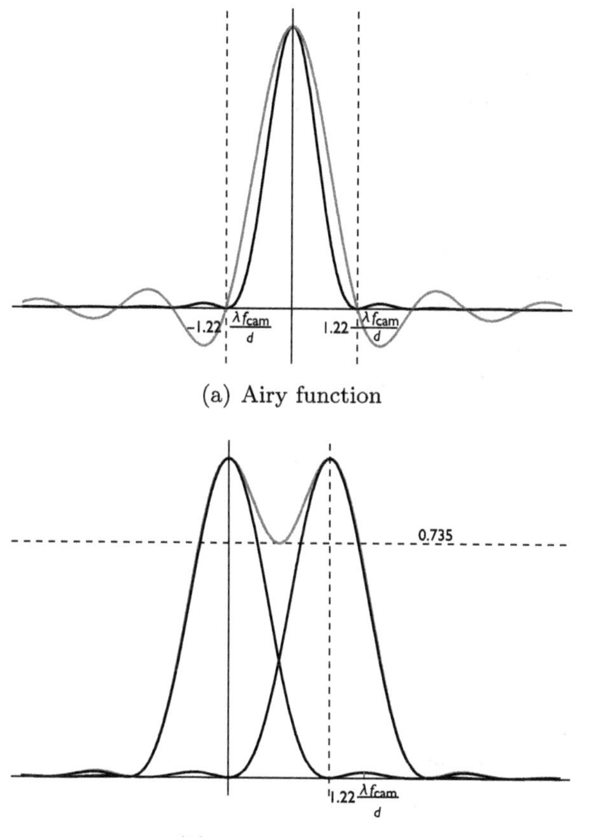

(a) Airy function

(b) Rayleigh criterion

Fig. 6.19. (a) The radial profile of the Fourier transform of a circular aperture (light grey), and the intensity of this (black) known as the Airy function. The first minima in intensity are at angles of $\pm 1.22\lambda/d$ or in the image plane at positions of $\pm 1.22\lambda f_{\mathrm{cam}}/d$. (b) The Rayleigh resolution criterion in which the two Airy patterns are separated by $1.22\lambda f_{\mathrm{cam}}/d$, such that the maximum of one overlaps with the first minimum of the other. The sum of the two Airy patterns at the Rayleigh criterion is shown by the grey line.

and $R_{\mathrm{slit}} = mN$, and therefore $R_{\mathrm{circ}} \approx R_{\mathrm{slit}}/1.22$. That is to say, the resolving power of a diffraction grating illuminated by a circular beam is less than that of a diffraction grating illuminated by a rectangular slit of the same width, by a factor 1.22, which is due to the fact that the circular beam does not illuminate the grating lines with the greatest separation as much as it illuminates the lines which are closer together.

6.4.2 *Single-mode fibre*

Now, let us consider the situation in which a spectrograph is fed with SMF, which is an important consideration for many astrophotonic applications in which diffraction limited performance is required, or in which the single-mode behaviour of waveguides are prerequisite (see Chapter 9, and especially Sections 9.1 and 9.2). In this case the diffraction grating is illuminated by the output of a single-mode fibre, which can be approximated by a Gaussian profile (Section 5.1.2.1) and for now we will assume that there is no truncation of the Gaussian profile. The PSF in the image plane is now given by the Fourier transform of a Gaussian, which is itself a Gaussian, which in the image plane is given by

$$E_{\mathrm{PSF}}(s) = E_0 \mathrm{e}^{-s^2}, \tag{6.31}$$

where

$$s = \frac{\pi w r}{\lambda f_{\mathrm{cam}}}, \tag{6.32}$$

cf. Equation 6.28, and w is the beam waist on the grating, i.e. the beam waist in the image plane is

$$w_0 = \frac{\lambda f_{\mathrm{cam}}}{\pi w}. \tag{6.33}$$

The Rayleigh criterion cannot be used with a Gaussian PSF, as there are no minima. Instead (still following Robertson and Bland–Hawthorn 2012), we define two Gaussian PSFs to be just resolved when separated by a distance Δx such that the sum of the two PSFs has a local minimum equal to that for two just-resolved Airy discs, i.e. a minimum equal to $0.735 \times$ the maximum, which occurs at a separation of $\Delta x \approx 1.415 w_0$, see Figure 6.20.

As before we can use this resolution criterion to calculate the resolving power using Equation 6.26,

$$R_{\mathrm{SMF}} = 2.22 \tan \theta_{\mathrm{i}} \frac{d_{\mathrm{SMF}}}{\lambda}, \tag{6.34}$$

where d_{SMF} is the beam diameter on the grating. Note, that this appears to give a higher resolving power than for a MMF. However, this is sensitive to the exact definition chosen for the resolution of two Gaussian PSFs. Different assumptions will yield a different leading factor. In any case, the illumination of the diffraction grating will not be a pure Gaussian but must necessarily be truncated at the edges, as will be discussed in the following section.

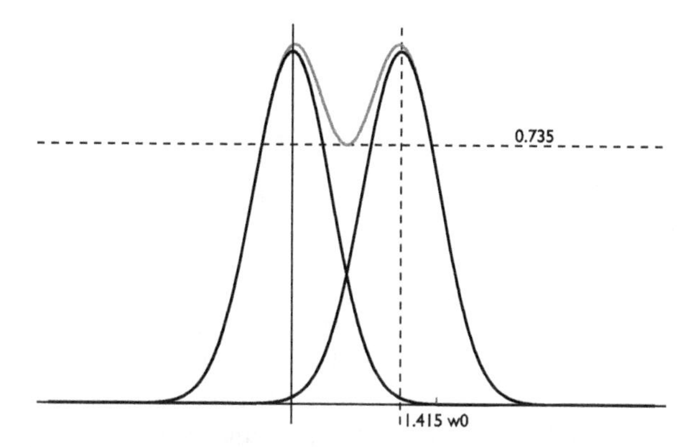

Fig. 6.20. The criterion used for the resolution of two Gaussian PSFs is taken to be when the local minimum between two PSFs is 0.735 of the maximum, as for the Rayleigh criterion of two resolved Airy functions, and occurs at a separation of $\approx 1.415 w_0$.

6.4.2.1 *Truncated Gaussian*

The above derivation for the resolving power of a single mode spectrograph can only be approximate since we assumed the PSF was given by the Fourier transform of a pure Gaussian. Since a Gaussian extends to infinity this cannot be true, and the output beam of the SMF must be truncated by some finite aperture optical element. Let us assume that the truncation is by a circular aperture, and define the truncation factor to be

$$T = \frac{d_{\rm SMF}}{D_{\rm opt}}, \tag{6.35}$$

where $D_{\rm opt}$ is the diameter of the limiting aperture.

The PSF of a truncated Gaussian in the image plane is again given by the Fourier transform, and is given by

$$E_{\rm trunc} = E_0 e^{-\frac{s^2\sigma^2}{2}} \frac{\operatorname{erf}\left(\frac{2-{\rm i}sT\sigma}{\sqrt{2}T}\right) + \operatorname{erf}\left(\frac{2+{\rm i}sT\sigma}{\sqrt{2}T}\right)}{2\pi\sqrt{2}}, \tag{6.36}$$

and the intensity is the absolute square of this. In the limit of $T >> 1$ this approaches an Airy function, and at $T << 1$ this approaches a pure Gaussian PSF, see Figure 6.21. It is possible to follow the analyses above to derive the resolving power due to a truncated Gaussian PSF for particular values of T and σ; we leave this as an exercise for the reader.

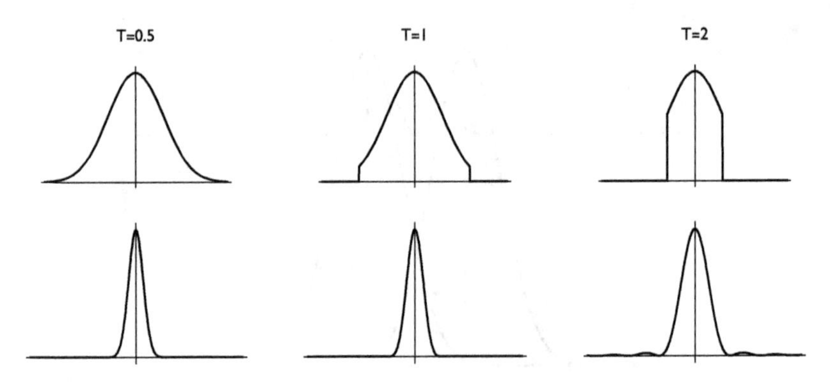

Fig. 6.21. A truncated Gaussian and the resulting PSF for different truncation factors.

6.4.3 *Spectrograph design*

Robertson and Bland–Hawthorn (2012) showed how to use the expressions given above to derive the basic parameters for a diffraction limited micro-spectrograph. For example, consider a micro-spectrograph fed by a single SMF with $NA = 0.11$, designed to have R = 20,000 operating at a wavelength of $\lambda_c = 1.5$ μm.

For a typical VPH grating angle of 30°, Equation 6.34 implies a size of $d_{\mathrm{SMF}} = 23.4$ mm. Working at first order, Equation 6.22 implies that this grating should have a line density of 667 lines mm^{-1} (i.e. $b = 1.5$ μm).

The beam exiting from the fibre has a focal ratio $F \sim 1/(2 * 0.11) = 4.5$. However, the optics must be somewhat faster than this in order to avoid a severe truncation of the Gaussian. A truncation of $T = 0.75$ is suggested by Robertson and Bland–Hawthorn (2012) as a compromise between avoiding this truncation without making the optics unnecessarily large and fast, yielding a focal ratio of $f/3.4$ for the collimator.

The camera lens (or lenses) must be chosen to give a reasonable size spot on the detector, i.e. one that is properly sampled (see Section 2.3). For a Gaussian PSF the beam diameter is $2w_0$, with w_0 given by Equation 6.33. Since we have $2w = d = 23.4$ mm at the grating, this leads to,

$$w_0 = \frac{2\lambda f_{\mathrm{cam}}}{\pi d_{\mathrm{SMF}}} = \frac{f_{\mathrm{cam}}}{24504} \text{ mm.} \tag{6.37}$$

Setting w_0 to be equal to 2.5 pixels, and using a pixel size of 15 μm gives a camera focal length of $f_{\mathrm{cam}} = 919$ mm.

In practice, these parameters would be derived using ray tracing software such as Zemax, but the examples given here serve to illustrate how the basic parameters can be derived from first principles. We again refer the reader to the full treatment given by Robertson and Bland–Hawthorn (2012) for further details.

6.4.4 *Coupling into channel waveguides*

In many astrophotonic applications, the waveguides are not simply used to transport light to the entrance slit of a spectrograph, but other photonic devices are used to manipulate the light. In this case, there is often the need to inject from one waveguide, typically a fibre, into another, typically a channel waveguide and getting the light into a single-mode fibre is only the beginning of the story. When high index contrast waveguides are used, such as silicon or silicon nitride, injection from a telescope into SMF is the easy part, and the injection from SMF into the extremely narrow channel waveguides is the real challenge, as will now be discussed.

Channel waveguides often have a high refractive index. For example, at $\lambda = 1.55$ μm, Si has $n \approx 3.4$, while Si_3N_4 has $n \approx 2.0$. Therefore, single-mode waveguides using these materials often have very narrow widths; for silicon-on-insulator (SOI) waveguides widths of 250–400 nm are common. A rule-of-thumb for single-mode waveguide dimensions is that they must be $< \lambda_0/n$, where λ_0 is the wavelength in vacuum.

Coupling light from a telescope to such waveguides presents an extra level of complication compared to feeding single-mode fibres. Even after coupling into single-mode fibres via a photonic lantern (see above), there is a severe mismatch in the mode field diameters of typical single-mode fibres and those of the SOI channel waveguides, so the overlap integral is very poor. For example, Figure 6.22 shows the modes of a typical SiO_2 single-mode fibre, and that of a Si waveguide with SiO_2 cladding and dimensions of 400×300 nm. The overlap integral between the two modes leads to losses of ≈ -13 dB.

6.4.4.1 *Inverted taper edge couplers*

One solution to this problem is to use an inverted taper on the waveguide, whereby the waveguide decreases in width at the edges of the chip, see Figure 6.23. As the width of the waveguide narrows, more of the light is squeezed out of the core into the evanescent field, exactly as for the photonic lantern (see Figure 6.14). Note that the waveguide need only be tapered in one dimension, *viz.* the width, in order to increase the

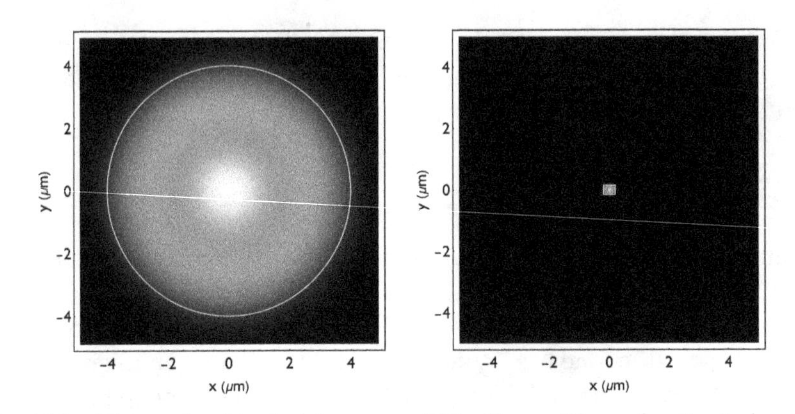

Fig. 6.22. The mode of typical 8 μm core SiO_2 fibre, with a refractive index difference of 0.01 (left) and that of a Si channel waveguide with SiO_2 cladding and dimensions of 400×300 nm. The mismatch in mode profiles results in difficulties when coupling single-mode fibres to SOI channel waveguides.

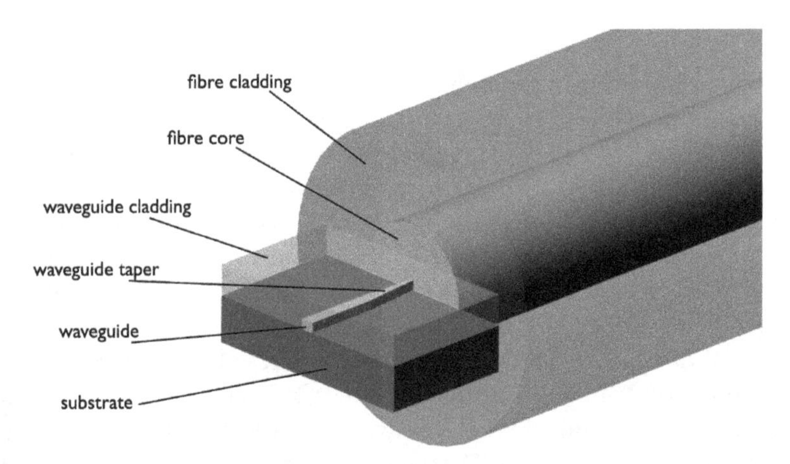

Fig. 6.23. Sketch of an inverted taper butt-coupled to a fibre. The dimensions are not drawn to scale, in order to make all components visible.

mode field diameter in this way. Figure 6.24 shows the coupled power between an 8 μm core fibre, and a Si_3N_4 waveguide in SiO_2 cladding, of height 650 nm and width 900 nm, which tapers over a distance 400 μm to a final tip width as plotted. The maximum coupled power occurs when the waveguide is tapered down to $\approx$50 nm. It is possible to achieve even higher coupling by introducing sub-wavelength grating structures into the waveguide. These are modulations with a pitch smaller than the

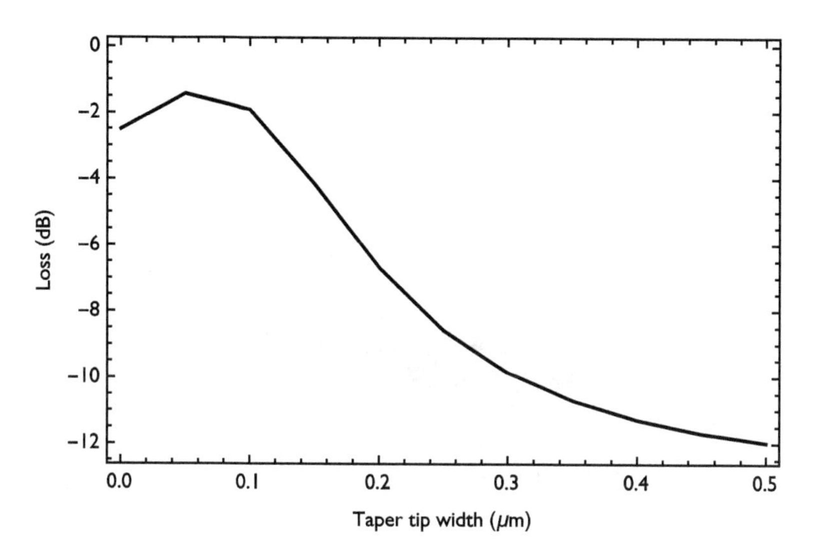

Fig. 6.24. The power coupled between a perfectly aligned 8 μm core fibre, and a Si_3N_4 waveguide in SiO_2 cladding, of height 650 nm and width 900 nm, which tapers over a distance 400 μm to a final tip width as plotted.

wavelength of the light being guided, which therefore act to modify the effective index of the waveguide, and the mode field diameter allowing for better matching of the modes (e.g. Halir *et al.* 2015).

Coupling between a fibre and an inverted waveguide taper may be further improved in several ways. For example, the tapered waveguide may be covered in a polymer cladding which acts as a new waveguide better matched to the fibre, and eases the alignment tolerances; the fibre itself may be lensed or of a high *NA* in order to match the fibre mode to that of the waveguide.

6.4.5 *Grating couplers*

A different fibre to chip coupling scheme is found in the grating coupler, as sketched in Figure 6.25. Here, a diffraction grating is written into the surface of the waveguide, which thus diffracts light of a specific wavelength at specific angles. A fibre can therefore be placed at the corresponding angle to the waveguide and thereby collect the diffracted light from the waveguide, or in reverse, feed light into the waveguide. Let us examine the simplified case shown in Figure 6.26.

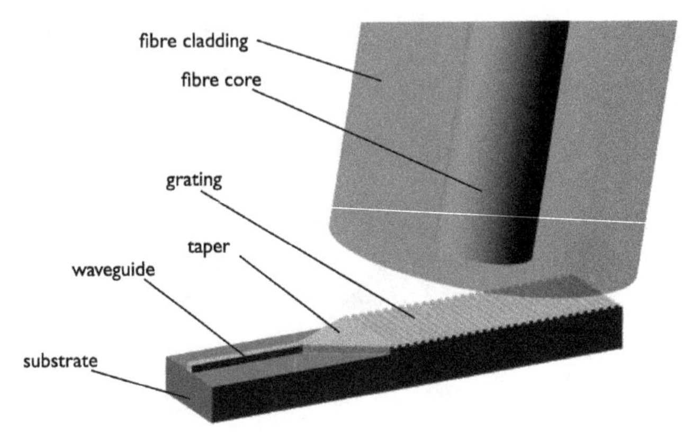

Fig. 6.25. Sketch of a grating coupler. The dimensions are not drawn to scale, in order to make all components visible.

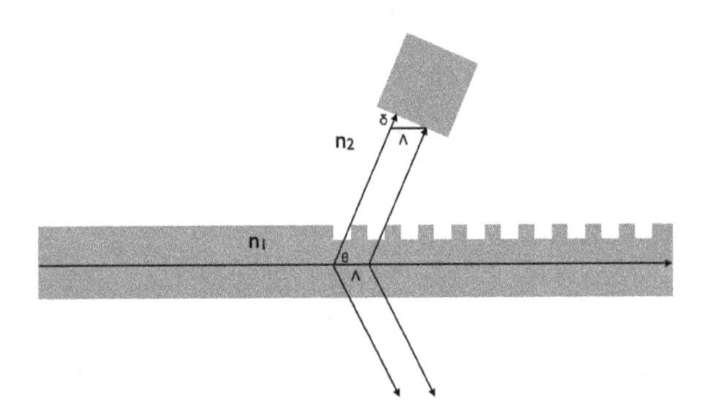

Fig. 6.26. Sketch of a simplified one-dimensional model of a grating coupler.

Consider a single-mode waveguide with light of wavelength λ propagating along its axis. Each tooth of the grating can be thought of as scattering the incident light in all directions; however, only at specific angles will the light from adjacent teeth add together in phase. If the grating period is Λ, then the phase difference between adjacent grating teeth for light scattered at an angle θ as in Figure 6.26 is

$$\begin{aligned} \text{phase difference} &= \frac{2\pi}{\lambda}(n_1\Lambda - n_2\delta)\,, \\ &= \frac{2\pi\Lambda}{\lambda}(n_1 - n_2\cos\theta)\,. \end{aligned} \tag{6.38}$$

When this light adds together in phase, we then have the condition

$$m\lambda = \Lambda\left(n_1 - n_2 \cos\theta_m\right), \tag{6.39}$$

where m is an integer. Thus, the grating will direct light at specific angles θ_m for a given order.

Note that in the simple case described above, the grating will not be very efficient. Light will be diffracted into many orders, and will also be diffracted downwards, away from the fibre, as well as backwards along the fibre. There are several strategies to increase the efficiency of grating couplers. The grating can be blazed to optimise the diffraction into a particular order. The grating can be detuned, such that each tooth is of a different width, although still at the same period Λ, to narrow the range of angles for each diffraction order. Also, reflectors may be placed along the bottom of the waveguide to minimise the light diffracted into the substrate.

Following such strategies grating couplers can reach peak efficiencies of > -1 dB. However, these efficiencies peak in a narrow range of wavelengths. Typical 3 dB bandwidths (the wavelength range over which the efficiencies are > -3 dB) are $\approx$40–70 nm. Compared to the typical astronomical passbands of $\sim$300 nm, grating couplers are very narrow band. They would be limited to particular niche astronomical applications in which only specific wavelengths are to be targeted, or else the full passband must first be split into a number of sub-windows.

6.5 Coupling between Waveguides

So far, we have considered coupling into waveguides via injection into the end-face. We now turn our attention to the coupling between waveguides. In the first case, we will consider the splitting of one waveguide into two, and vice versa. Second, we consider evanescent coupling of adjacent waveguides. Some of the mathematics in this section is more advanced than most of the other examples in this book, and requires a knowledge of linear algebra and eigenvalues. Furthermore, to avoid very lengthy explanations we have only given very truncated derivations. If necessary, one can skip directly to the results.

6.5.1 *Y branch splitters and beam combiners*

Before looking at coupling between two separate waveguides, we will first consider the case in which one waveguide is split into two, or vice versa. Consider Figure 6.27 in which a single-mode waveguide is split into two single-mode waveguides identical to the first, via a tapered coupling region. In this case, the light from the input waveguide is split equally into the two outgoing waveguides, and thus acts as a 3 dB splitter.

In the reverse situation of two single-mode waveguides combining into one, as in Figure 6.28, the two input waveguides will not necessarily couple all their power to the output waveguide: in general the coupling will be lossy. The time-reversed analogue of Figure 6.27, in which all the power is coupled into a single outgoing waveguide, is possible only when the two inputs are exactly coherent (Figure 6.28a). If the light in the two input waveguides is exactly out of phase then no light will couple to the outgoing waveguide; it will all be radiated and lost (Figure 6.28b).

The general situation can be understood by considering that the two incoming modes will excite exactly two normal modes (see Section 6.5.2 for a discussion of normal modes) in the tapered combining section (Figure 6.28c). However, the outgoing waveguide can only carry one mode, *viz.* the even normal mode; the higher order odd normal mode will be lost. The final power will therefore depend on the relative phase of the incoming modes, or equivalently on the relative power in the even and odd modes.

Interestingly, if light is input into only one of the incoming waveguides, then both even and odd normal modes are equally excited, and the coupled power can not exceed 3 dB for a symmetric coupler.

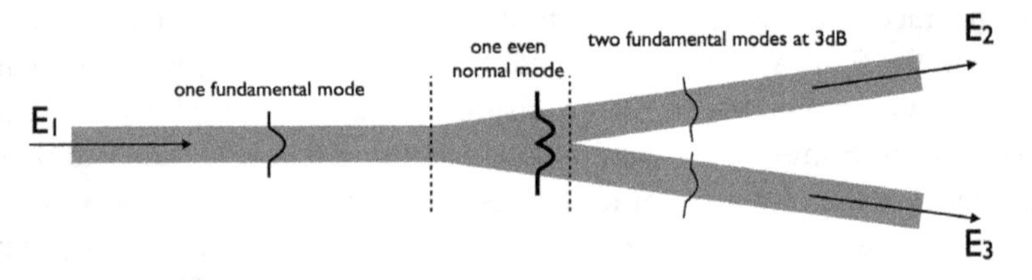

Fig. 6.27. Schematic diagram of a 3 dB beam splitter, consisting of one single-mode waveguide splitting into two. Each of the outgoing modes carries 50% of the incoming power.

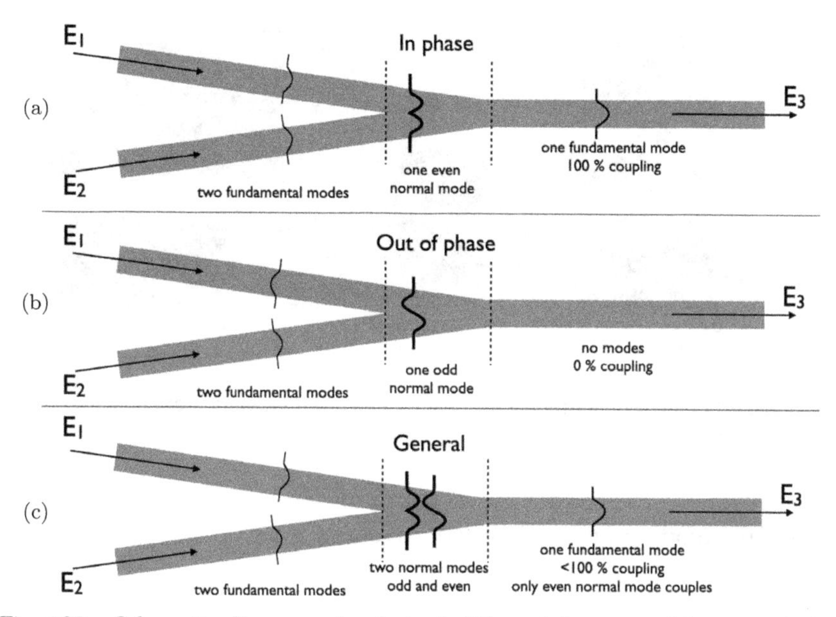

Fig. 6.28. Schematic diagram of a photonic Y branch beam combiner, consisting of two single-mode waveguides combining into one. The power in the outgoing mode depends on the relative phase of the incoming modes.

6.5.2 *Coupling between waveguides*

When two waveguides are close enough, the evanescent field from the first can overlap with the core of the second. In this way, it is possible to couple light from one waveguide to another, as shown in Figure 6.29.

We will derive the behaviour of coupled waveguides using **normal mode analysis.** Normal mode analysis can be applied to any arbitrary system of waveguides, but for a concrete example, let us consider a system of N single-mode waveguides. The normal modes of this system, also referred to as **supermodes,** are sets of particular amplitude and phase relationships for the modes of each core, for which there is no power transfer between the cores. N.B. the normal modes are modes of the entire system, not of the individual cores, that is they are sums of the individual modes of each core with the particular amplitude and phase relationships such that there is no power transfer between the cores. For a system of N cores there are N normal modes, although these are not all necessarily non-degenerate.

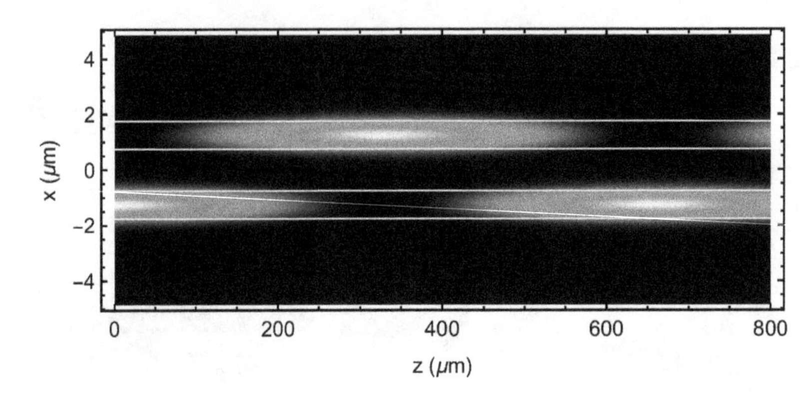

Fig. 6.29. The power coupled between two parallel waveguides as a function of distance along the waveguides.

In general, we can write the modes of the composite waveguide as the sum of the modes of the individual waveguides

$$\Psi = \sum_j a_j \widehat{\psi}_j, \tag{6.40}$$

where ψ_j are the modes of the individual waveguides and a_j are the amplitudes of each mode. The scalar wave equation of each of the individual waveguide modes is given by

$$\nabla^2 \widehat{\psi}_j + (k^2 n_j^2 - \beta_j^2)\widehat{\psi}_j = 0, \tag{6.41}$$

where β_j are the propagation constants of each mode, and n_j is the refractive index of each core, which can be treated as a perturbation of the background index

$$n_j^2(x, y) = n_0^2 + \Delta n_j^2(x, y), \tag{6.42}$$

$$\Delta n_j^2(x, y) = \begin{cases} n_{\text{co}}^2 - n_0^2, & \text{in each core } j, \\ 0, & \text{elsewhere.} \end{cases} \tag{6.43}$$

Similarly, the scalar wave equation for the normal modes is given by

$$\nabla^2 \Psi + (k^2 n^2 - \beta^2)\Psi = 0, \tag{6.44}$$

where

$$n^2(x, y) = n_0^2 + \sum_{j=1}^{N} \Delta n_j^2(x, y). \tag{6.45}$$

Furthermore, Ψ is independent of z by definition, since for normal modes, the individual waveguide modes $a_j\widehat{\psi}_j$ add together, such that no power is transferred between them.

In the case of weak guidance, i.e. when the refractive index contrast between the core and the cladding is small, Equations 6.40–6.45 can be shown to lead to

$$\Delta\beta_m a_m - \sum_{j \neq m} C_{jm} a_j = 0, \tag{6.46}$$

$$C_{jm} = \frac{k}{2n_0} \int_A \widehat{\psi}_j \widehat{\psi}_m \Delta n_m^2 \, dA, \tag{6.47}$$

where the C_{jm} are the coupling coefficients, and

$$\Delta\beta_m = (\beta - \beta_m), \tag{6.48}$$

is the difference between the propagation constant of the normal modes and the propagation constant of the modes of the individual waveguides. So we have N simultaneous equations for a_m, one for each mode, and these give the modes which add together to form a normal mode.

Example 6.5: Find the normal modes for a 2×2 directional coupler, such as that shown in Figure 6.29.

In this case the propagation constants of the individual waveguides are identical, $\beta_1 = \beta_2 = \beta_0$, and therefore $\Delta\beta = \beta - \beta_0$ for all of the normal modes. Furthermore, the coupling between waveguides is identical, so Equation 6.46 becomes

$$\begin{pmatrix} \Delta\beta & -C \\ -C & \Delta\beta \end{pmatrix} \begin{pmatrix} a_1 \\ a_2 \end{pmatrix} = \begin{pmatrix} 0 \\ 0 \end{pmatrix}. \tag{6.49}$$

Now, because we have a matrix equation of the form $M \cdot A = 0$ it follows that $|M| = 0$. Therefore, to find the $\Delta\beta$ we can find the determinant, set it equal to zero, and solve for $\Delta\beta$. This equation will have $N = 2$ roots. The normal mode Ψ is given by the eigenvectors A corresponding to each $\Delta\beta$. Doing so yields

$$\Delta\beta = \pm C. \tag{6.50}$$

(*Note*: $\sum \Delta\beta = 0$, as it must.)

The corresponding eigenvectors can now be found by solving

$$\begin{pmatrix} C & -C \\ -C & C \end{pmatrix}\begin{pmatrix} a_1 \\ a_2 \end{pmatrix} = \begin{pmatrix} 0 \\ 0 \end{pmatrix}, \begin{pmatrix} -C & -C \\ -C & -C \end{pmatrix}\begin{pmatrix} a_1 \\ a_2 \end{pmatrix} = \begin{pmatrix} 0 \\ 0 \end{pmatrix}, \tag{6.51}$$

which gives,

$$\widehat{\Psi}_1 = \frac{1}{\sqrt{2}}\left(\widehat{\psi}_1 + \widehat{\psi}_2\right), \tag{6.52}$$

$$\widehat{\Psi}_2 = \frac{1}{\sqrt{2}}\left(\widehat{\psi}_1 - \widehat{\psi}_2\right). \tag{6.53}$$

These normal modes are orthogonal, and are therefore independent.

Finally, let us use the normal modes to find the individual modes of the waveguides. Let us assume that at $z = 0$ all the power is in the first waveguide, therefore

$$\widehat{\Psi}(0) = \widehat{\psi}_1. \tag{6.54}$$

We can now re-write this in terms of normal modes, for which we have calculated the propagation constants

$$\widehat{\Psi}(0) = \frac{1}{\sqrt{2}}\left(\widehat{\Psi}_1 + \widehat{\Psi}_2\right). \tag{6.55}$$

And letting each mode propagate independently

$$\begin{aligned} \widehat{\Psi}(z) &= \frac{1}{\sqrt{2}}\left(\widehat{\Psi}_1 e^{i(\beta_0+\Delta\beta_1)z} + \widehat{\Psi}_2 e^{i(\beta_0+\Delta\beta_2)z}\right), \\ &= e^{i\beta_0 z}\left(\widehat{\psi}_1 \cos Cz + \widehat{\psi}_2 i \sin Cz\right). \end{aligned} \tag{6.56}$$

That is to say the amplitudes in each core, a_1 and a_2 are

$$\begin{aligned} a_1 &= e^{i\beta_0 z} \cos Cz, \\ a_2 &= i e^{i\beta_0 z} \sin Cz. \end{aligned} \tag{6.57}$$

Note that in Equation 6.57 the amplitude in core two is $\pi/2$ out of phase with core 1, since $e^{i\frac{\pi}{2}} = i$. Therefore, when light couples between one waveguide and another there is always a $\pi/2$ phase change; if light couples from core 1 to core 2 and back to core 1 it will be out of phase with the original injected light. The power in each core is

$$P_1 = \cos^2 Cz, \tag{6.58}$$

$$P_2 = \sin^2 Cz. \tag{6.59}$$

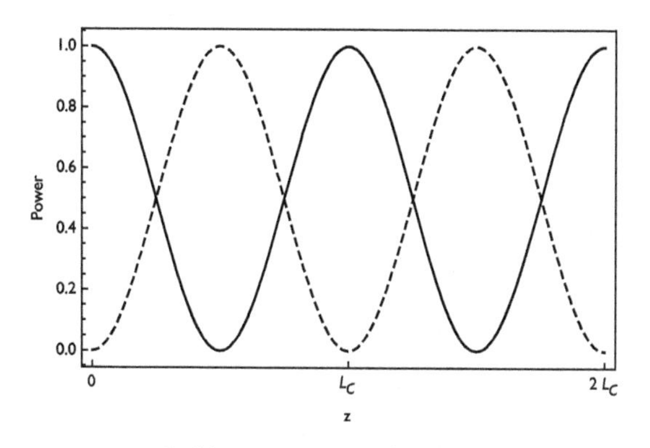

Fig. 6.30. The power coupled between two adjacent waveguides as a function of distance, if all the power is initially injected into one waveguide.

So the power oscillates between the two cores, with a beat length

$$L_C = \frac{\pi}{C}, \tag{6.60}$$

as shown in Figure 6.30.

Note that by controlling the length of the coupling region, any fraction of power can be coupled into the second waveguide.

6.6 Photonic Lanterns

Normal mode analysis can also provide a useful insight into the behaviour of photonic lanterns. The starting point are the simultaneous equations 6.46, and the coupling coefficients in Equation 6.47. To make the problem tractable, we will consider the simplified case of a 1×7 photonic lantern, arranged in a hexagonal packing.

The coupling coefficient between cores of the photonic lantern will in general depend on their distance, as sketched in Figure 6.31. Thus, the simultaneous equations governing the amplitudes of each normal mode become,

$$\begin{pmatrix} \Delta\beta & -C_1 & -C_1 & -C_1 & -C_1 & -C_1 & -C_1 \\ -C_1 & \Delta\beta & -C_1 & -C_2 & -C_3 & -C_2 & -C_1 \\ -C_1 & -C_1 & \Delta\beta & -C_1 & -C_2 & -C_3 & -C_2 \\ -C_1 & -C_2 & -C_1 & \Delta\beta & -C_1 & -C_2 & -C_3 \\ -C_1 & -C_3 & -C_2 & -C_1 & \Delta\beta & -C_1 & -C_2 \\ -C_1 & -C_2 & -C_3 & -C_2 & -C_1 & \Delta\beta & -C_1 \\ -C_1 & -C_1 & -C_2 & -C_3 & -C_2 & -C_1 & \Delta\beta \end{pmatrix} \begin{pmatrix} a_1 \\ a_2 \\ a_3 \\ a_4 \\ a_5 \\ a_6 \\ a_7 \end{pmatrix} = \begin{pmatrix} 0 \\ 0 \\ 0 \\ 0 \\ 0 \\ 0 \\ 0 \end{pmatrix}, \tag{6.61}$$

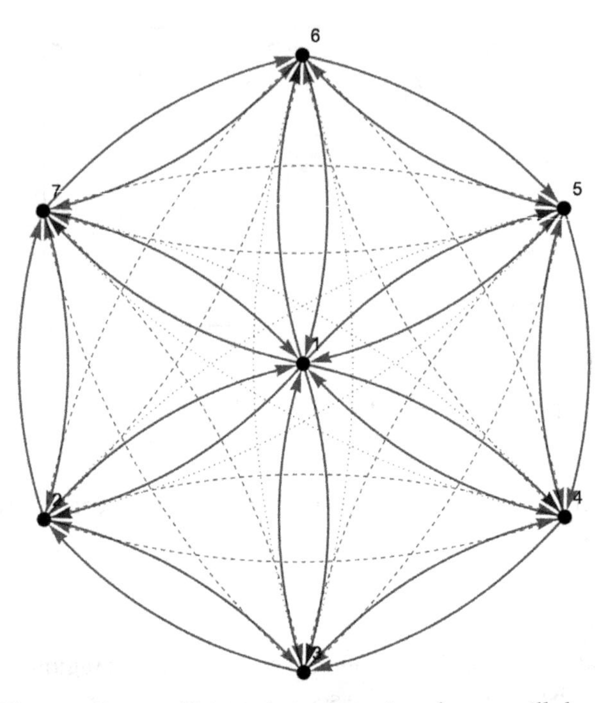

Fig. 6.31. The coupling coefficients between pairs of cores will depend on their separation. Nearest neighbours are shown with thick lines, second nearest neighbours with dashed lines, and third nearest neighbours with dotted lines. These correspond to the matrix shown in Equation 6.61.

where C_i is the coupling coefficient for the ith nearest neighbour. We can find the amplitudes a_i, and hence the normal modes, following the same procedure as for the directional coupler above, that is we find the determinants of the coupling matrix, solve to find $\Delta\beta$, then substitute the $\Delta\beta$s back in to get the corresponding eigenvectors. Doing so yields

$$\widehat{\Psi}_1 = \frac{1}{\sqrt{4}}\left(\widehat{\psi}_2 + \widehat{\psi}_3 - \widehat{\psi}_5 - \widehat{\psi}_6\right), \tag{6.62}$$

$$\Psi_2 = \frac{1}{\sqrt{8}}\left(\widehat{\psi}_2 - \widehat{\psi}_3 - 2\widehat{\psi}_4 - \widehat{\psi}_5 + \widehat{\psi}_6 + 2\widehat{\psi}_7\right), \tag{6.63}$$

$$\Psi_3 = \frac{1}{\sqrt{6}}\left(\widehat{\psi}_2 - \widehat{\psi}_3 + \widehat{\psi}_4 - \widehat{\psi}_5 + \widehat{\psi}_6 - \widehat{\psi}_7\right), \tag{6.64}$$

$$\Psi_4 = \frac{1}{\sqrt{8}}\left(\widehat{\psi}_2 + \widehat{\psi}_3 - 2\widehat{\psi}_4 + \widehat{\psi}_5 + \widehat{\psi}_6 - 2\widehat{\psi}_7\right), \tag{6.65}$$

$$\Psi_5 = \frac{1}{\sqrt{4}}\left(\widehat{\psi}_2 - \widehat{\psi}_3 + \widehat{\psi}_5 - \widehat{\psi}_6\right), \tag{6.66}$$

$$\Psi_6 = \frac{1}{\sqrt{1+6K_1}}\left(\widehat{\psi}_1 + K_1\left(\widehat{\psi}_2 + \widehat{\psi}_3 + \widehat{\psi}_4 + \widehat{\psi}_5 + \widehat{\psi}_6 + \widehat{\psi}_7\right)\right), \tag{6.67}$$

$$\Psi_7 = \frac{1}{\sqrt{1+6K_2}}\left(\widehat{\psi}_1 + K_2\left(\widehat{\psi}_2 + \widehat{\psi}_3 + \widehat{\psi}_4 + \widehat{\psi}_5 + \widehat{\psi}_6 + \widehat{\psi}_7\right)\right), \tag{6.68}$$

$$K_1 = \frac{2C_1 + 2C_2 + C_3 - \sqrt{24C_1^2 + (2C_1 + 2C_2 + C_3)^2}}{12C_1}, \tag{6.69}$$

$$K_2 = \frac{2C_1 + 2C_2 + C_3 + \sqrt{24C_1^2 + (2C_1 + 2C_2 + C_3)^2}}{12C_1}. \tag{6.70}$$

All these normal modes are orthogonal, i.e. $\Psi_i.\Psi_j = \delta_{ij}$.

What the above equations tell us is that the seven degenerate modes of the individual single-mode fibres, evolve into seven independent non-degenerate modes of the multimode fibre as the cores couple to one another; each of the seven single-modes of the fibre evolves into a separate mode despite the symmetry of the device.

We will not compute the modes of the photonic lantern further. Doing so would not be useful, since in general there will be a random initial phase of each mode due the seeing and therefore the coupling of the out-of-phase modes will be essentially random.

PART II

ASTROPHOTONIC DEVICES

In the first part of this book, we have developed the principles and concepts necessary for understanding astrophotonic instruments. We now use those principles to describe the astrophotonic instruments themselves. We do not review the entire field, nor do we delve into the details of the instrument design, rather our focus is again to make clear the principles by which the instruments work. We have organised the instruments based first on their astronomical function, and thereafter by the photonic components of the instruments. In this part, we focus only on those instruments which have currently reached a significant stage of development, having been either tested in the laboratory, in prototype instruments or in fully developed on-telescope instruments. In the next part, we focus on future planned and possible developments in astrophotonics.

CHAPTER 7

INTERFEROMETRY

One of the first major successes of astrophotonics has been in optical interferometry. Photonic components have played a transformative role in this field and they are crucial elements in state-of-the-art optical interferometers (Norris 2018). In this application the light from two or more separate telescopes is coherently combined to record interference due to the phase difference from the different path lengths from the source to the telescopes, by which means very high angular resolution observations can be obtained (see Section 3.5).

The application of photonics to optical interferometry dates back to the suggestion of using single-mode fibres to transport the light from telescopes to the beam combiners by Froehly (1981). Today the main area of application is in the beam combination itself.

7.1 Fibre Combiners

The first astrophotonic beam combiners used single-mode fibres to combine the beams from two telescopes. An outline of the scheme is shown in Figure 7.1, following Coudé du Foresto *et al.* (1997).

The fibre beam combiner takes light from each of the input fibres and splits it equally into the two output fibres, such that each output fibre contains 50% of the light from each telescope which interfere. By varying the optical path length in one arm, and recording the output intensity, fringes can be recorded, and from these the visibility can be computed (see Equation 3.40).

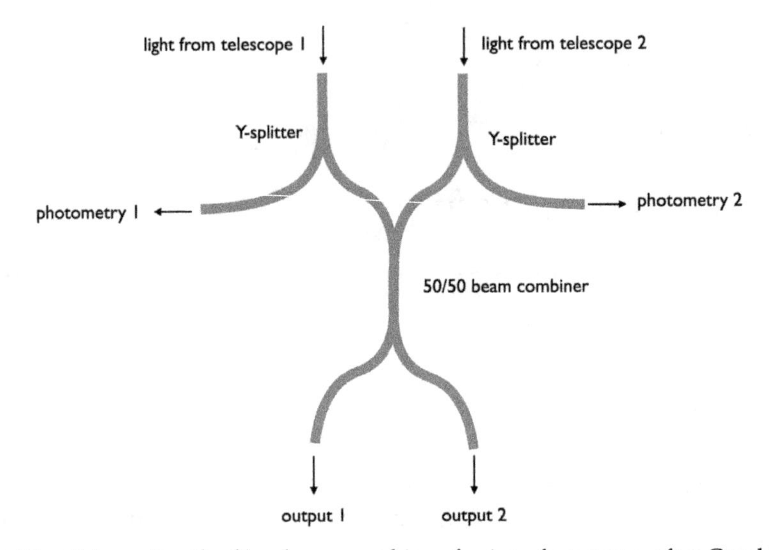

Fig. 7.1. Schematic of a fibre beam combiner for interferometry, after Coudé du Foresto *et al.* (1997). Light from the two telescopes is injected into single-mode fibres. A Y-splitter on each is used to siphon a small fraction of light to record the intensity in each fibre. The two fibres are then coupled together, and the interference is recorded at the two outputs. See the description in Section 7.1 for details.

The combiner works via the principle of evanescent coupling described in Chapter 6, in which the evanescent field from one fibre overlaps with the core of the adjacent fibre, and begins to be guided by it. A fibre beam combiner is constructed by twisting two fibres together, and then fusing them together.

7.1.1 *Spatial filtering*

The real advantage of using fibres to accomplish the beam combination is the spatial filtering provided by single-mode fibres. Consider coupling the beams from two telescopes together using any general method. The atmospheric turbulence above each telescope will destroy any coherence between the two telescopes, and recovering visibilities will have to rely on statistical calibration methods.

In contrast, all spatial information is lost when coupling light into a single-mode fibre; there is only one mode, and the transverse electric field distribution is governed by the wavelength of the light and the properties of the fibre alone (Section 5.1.2). Therefore, the changing seeing will not change the coherence of light coupled via single-mode waveguides.

However, the coupling efficiency into the SMF will change with as the seeing changes, since the overlap integral between the incoming electric field and the fibre mode will be changing (Chapter 6). However, by introducing a Y-splitter (Section 6.5.1), as in Figure 7.1, and recording the photometry from this channel, the changes in coupling efficiency can be easily calibrated, and the visibilities can be recovered.

Spatial filtering with single-mode waveguides has led to an order of magnitude improvement in the ability to recover visibility fringes, see Figure 7.2. The advantage of using single-mode waveguides to perform beam combination was first demonstrated using fibre cross-couplers

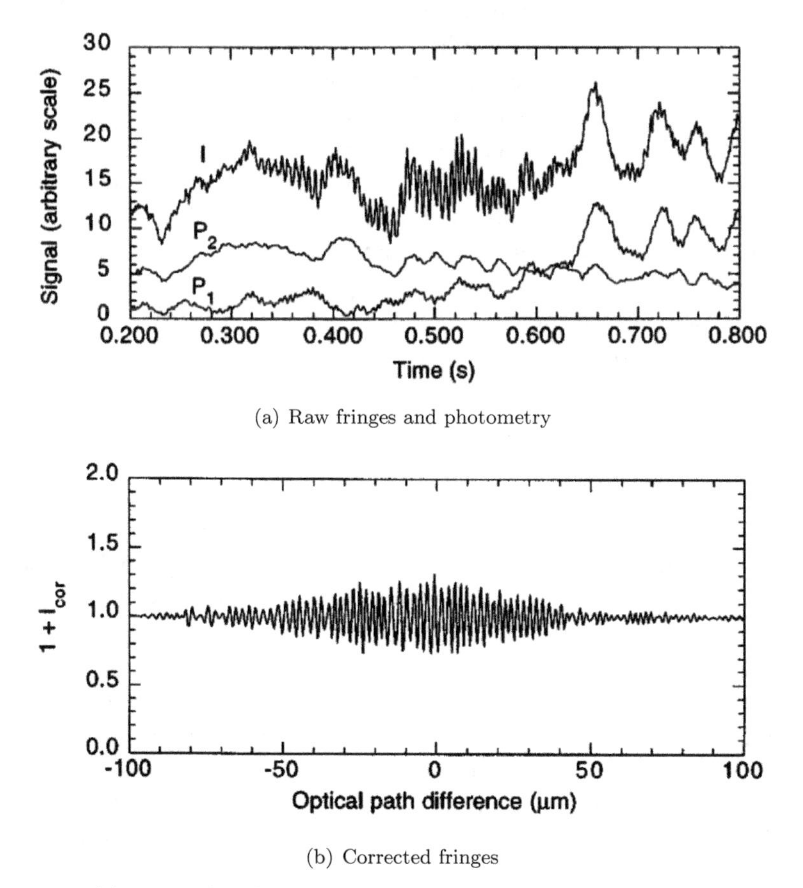

(a) Raw fringes and photometry

(b) Corrected fringes

Fig. 7.2. (a) Example of raw recorded fringes and photometry and (b) corrected fringes. Reproduced with permission from Coudé du Foresto *et al.* (1997). © European Southern Observatory.

(Coudé du Foresto and Ridgway 1992) on the FLUOR instrument on the IOTA telescope, which used ZBLAN fluoride fibre (see Section 5.1.1).

7.2 Integrated Photonic Beam Combiners

The idea of single-mode waveguide beam combiners was further refined with the introduction of integrated photonic beam combiners (Berger *et al.* 2001; Kern *et al.* 1997). Now, rather than using SMF as the combiners, lithographic waveguides on a planar chip are used instead. This has several advantages. First, the length of the waveguides can be more precisely tuned and maintained, since the waveguides are rigid. Second, the beam combination and Y-splitters can also be tuned more accurately, as this is done during the lithographic production of the waveguides, rather than after. Third, it is easier to extend the beam combination to include more base lines, as sketched in Figure 7.3. Fourth, there is only one interference output per baseline, increasing signal-to-noise. The scheme of an integrated beam combiner is shown in Figure 7.3, first repeating the two telescope scheme of Figure 7.1, and then extending the idea to three telescopes.

In the schemes of Figures 7.1 and 7.3, the path length must still be modulated in order to produce the fringes. However, integrated photonic circuits allow the introduction of phase shifting beam splitters in the waveguides, such that each beam combiner can sample the interference at different phases, e.g. 0, $\frac{\pi}{2}$, π, and $\frac{3\pi}{2}$ as in the lower left diagram of Figure 7.4. This technique is sometimes known as ABCD beam combination. This leads to the possibility of having complex integrated beam combiners, such as that shown in the main diagram of Figure 7.4, capable of recording the fringes from each combination of four telescopes, without the need for optical path length modulation.

The state-of-the-art of photonic beam combination can be found in the European Southern Observatory's GRAVITY instrument on the VLT (Gravity Collaboration *et al.* 2017). This combines the light from 4×8 m telescopes using a phosphor-doped silica on a silicon wafer manufactured by the Laboratoire d'Électronique des Technologies de l'Information (CEA/LETI) using plasma-enhanced chemical vapour deposition (Jocou *et al.* 2014).

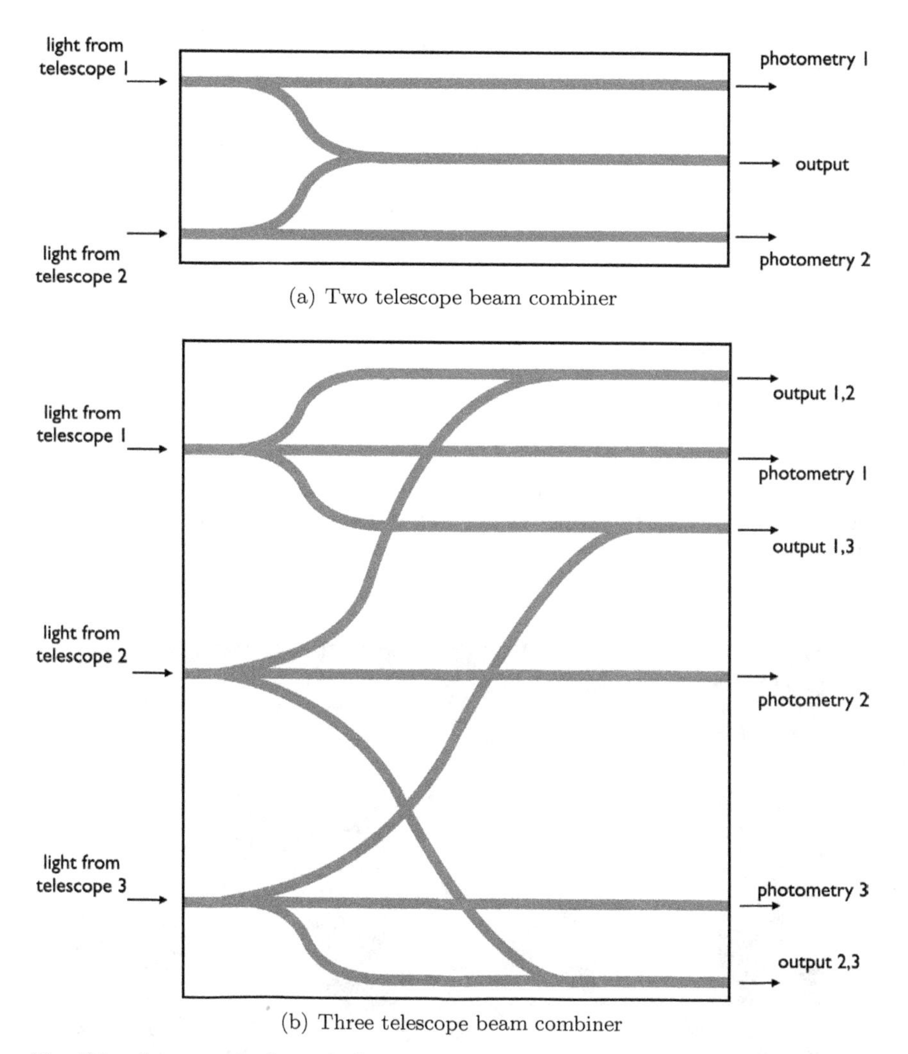

(a) Two telescope beam combiner

(b) Three telescope beam combiner

Fig. 7.3. Integrated photonic beam combiner schemes for two and three telescopes.

GRAVITY has tested the equivalence principle of general relativity in hitherto untested regimes through observations of stellar orbits around the super massive black hole at the centre of our Galaxy (Gravity Collaboration *et al.* 2019), and has made the most detailed observations of material

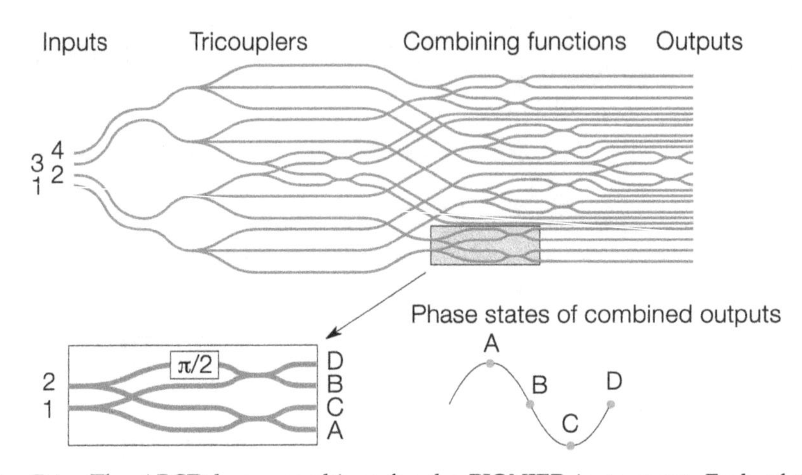

Fig. 7.4. The ABCD beam combiner for the PIONIER instrument. Each of the beams from the four telescopes of the VLT is combined with the others at four discrete phases, allowing the phases to be recovered from the photometry of each output. Reproduced with permission from Zins *et al.* (2011).

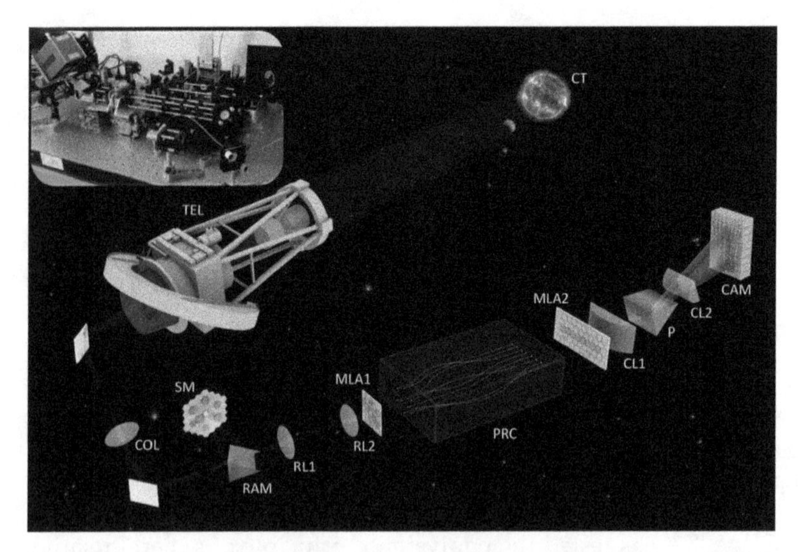

Fig. 7.5. Schematic of the aperture masking Dragonfly instrument. Selected waveguides in the pupil remapper (labelled PRC) can be fed using the MEMS micro mirror device (marked SM) to provide non-redundant interference at the output. Reproduced with permission from Jovanovic *et al.* (2012b). © Oxford University Press.

orbiting close to a black hole, measuring velocities of $\approx$30% the speed of light (Gravity Collaboration *et al.* 2018). These unprecedented observations are a testament to the power of photonics in astronomical interferometry.

7.3 Aperture Masking Interferometry

Another technique in interferometry in which photonics is playing a transformative role is aperture masking. Here, rather than combining the beam from several telescopes, the telescope pupil is masked to produce several small sub-apertures, which are then interferometrically combined to reproduce the full diffraction limited resolution of the telescope, which is usually unobtainable due to the distortion of the incoming wavefronts due to atmospheric turbulence across the entrance pupil.

Whilst this could be (and indeed has been) achieved by physically masking the entrance pupil of the telescope (Tuthill 2018), photonics allows the masking to be combined with the beam combination. Direct-write waveguides (Section 5.5) are used to segment a re-imaged telescope pupil and remap this into a series of non-redundant baselines, which can be combined to produce the interference (Jovanovic *et al.* 2012b; Norris *et al.* 2014). The direct-write remapping waveguides can be fed into a photonic beam combiner, such as that described above to exploit the advantages already described, including the phase-shifting beam combination to recover fringes (Norris *et al.* 2015). Figure 7.5 shows a schematic of the Dragonfly instrument (Jovanovic *et al.* 2012b).

CHAPTER 8

OH SUPPRESSION

In this chapter, we discuss the use of photonic filters to suppress emission from the Earth's atmosphere in order to reduce the background of astronomical observations, and thereby increase the sensitivity of the observations.

The night-sky is extremely bright at near-infrared wavelengths, with a surface brightness at a wavelength of 1.6 μm of $\approx$14 mag arcsec^{-2}, compared to $\approx$20 mag arcsec^{-2} at 0.55 μm, i.e. 250 times brighter; see Figure 8.1. This background is much brighter than typical sources of interest, and therefore leads to high Poissonian noise and low signal-to-noise (Equation 3.1). Furthermore, the background varies rapidly, both temporally and spatially, and accurate background subtraction is often marred by systematic errors.

Almost all of the background between 1 and 1.8 μm is due to the rovibrational decay of OH molecules in an $\approx$11 km thick layer at an altitude of $\approx$87 km, which gives rise to an extremely bright forest of emission lines, as shown in Figure 8.2.

8.1 The Idea of OH Suppression

Although, the OH lines are extremely bright and variable, they are also extremely narrow (full width at half-maximum, FWHM, of $\approx 5 \times 10^{-12}$ m), and occur in doublets separated by $\sim 10^{-10}$ m, and the surface brightness between them is very dark ($\approx$300–600 ph s^{-1} m^{-2} μm^{-1} arcsec^{-2}; see Figure 8.2). Thus, if the OH lines can be selectively removed, whilst leaving the rest of the spectrum intact, the near-infrared night sky would be

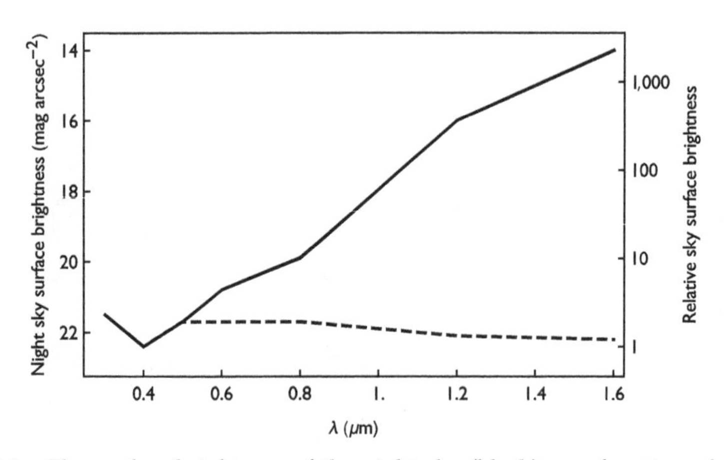

Fig. 8.1. The surface brightness of the night sky (black) as a function of wavelength. If there were no OH emission the relative brightness would be as shown by the dashed line.

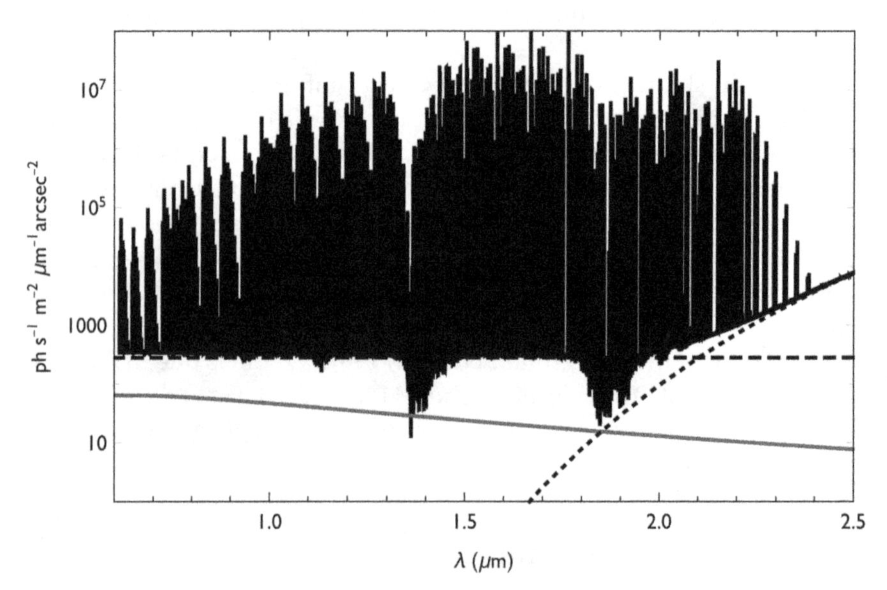

Fig. 8.2. The background emission for a ground-based telescope. The total background is shown in black, and is dominated by a forest of emission lines, 80% of which are from atmospheric OH molecules. The continuum between these lines is comprised of the zodiacal scattered light (grey), thermal emission from the telescope (dotted) and an unidentified interline continuum (dashed). (For full details on the NIR background see e.g. Abrams *et al.* 1994; Content 1996; Rousselot *et al.* 2000; Ellis and Bland-Hawthorn 2008; Sullivan and Simcoe 2012; Trinh *et al.* 2013a; Oliva *et al.* 2015; Nguyen *et al.* 2016 and references therein).

very dark, darker in fact than at visible wavelengths. This is the idea of OH suppression.

At first sight it may seem that this ought to be trivial: simply observe at a high spectral resolving power and then remove the OH lines *post facto* with the data processing software. Unfortunately, this scheme cannot realise the full benefit of OH suppression. This is because all spectrographs unavoidably scatter light due to imperfections in the surface quality of lenses, mirrors and in the diffraction gratings. Thus, the light from a particular emission line will be spread out over a number of pixels. For example, Figure 8.3 shows the point spread function (PSF) from the KOALA IFU and AAOmega spectrograph (Ellis *et al.* 2014).

The upshot of this is illustrated in Figure 8.4, which models the smearing of the OH lines due to spectrograph scattering. If the OH lines are subsequently removed in software, only the cores of the lines will be removed, whereas the interline regions will still be contaminated by the residual OH emission. It is therefore necessary either to remove the OH light before it enters the spectrograph, or else to remove it selectively based on the wavelength, or else to very accurately control the effects of scattering.

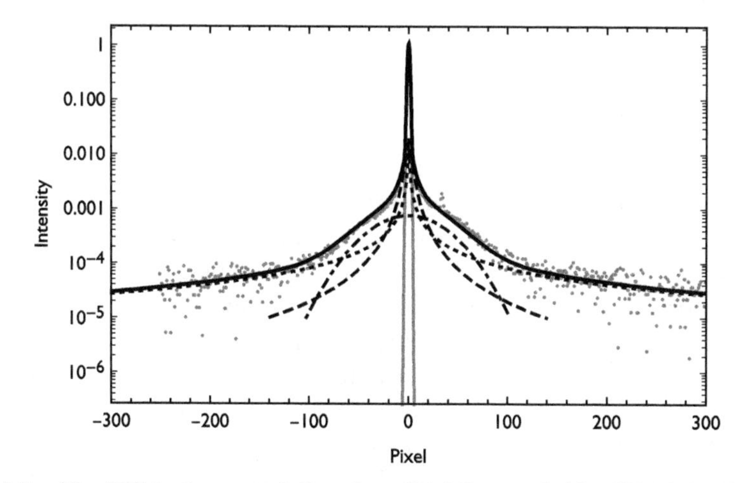

Fig. 8.3. The PSF in the spatial direction of AAOmega fed by KOALA. The PSF (thick black line) can be described by the sum of a Gaussian due to the slit width (grey line), a Lorentzian due to scattering from surface roughness etc. (dashed line), a $1/r$ term due to bulk scattering in the optics (dotted line) and a faint broad Gaussian due to aberrations in the spectrograph or ghosts (dot-dashed line). Additionally, in the spectral direction there will be a weaker Lorentzian term due to diffraction (Woods *et al.* 1994).

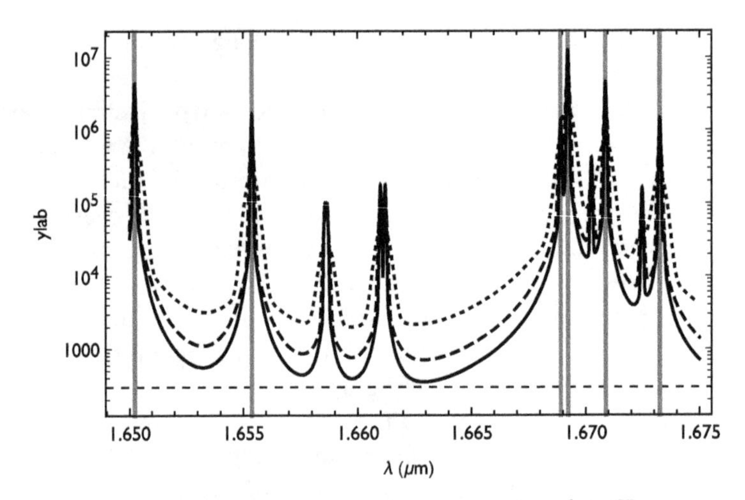

Fig. 8.4. The spectrum of the OH lines between 1.65 and 1.675 μm convolved with the spectrograph point spread function, at resolving powers of 3,000 (dotted), 10,000 (dashed) and 20,000 (continuous). The interline region is contaminated by the Lorentzian scattering wings of the PSF. If the OH lines are subsequently masked at high resolution, e.g. the grey lines, only the cores of the lines will be removed, but the interline region will still be polluted with OH emission. The dashed line shows the measured interline continuum; it is possible that this may be partly due to scattered OH emission.

8.2 OH Suppression with Fibre Bragg Gratings

The requirements for OH suppression are, *prima facie*, rather daunting. One must simultaneously attain a series of very narrow filters at specific, but irregular wavelengths, whilst maintaining high throughput at all other wavelengths, and one must achieve this without inadvertently scattering the light or introducing other unwanted aberrations.

These requirements are extremely difficult to achieve with conventional filters. The large number of notches needed requires complicated filters with many layers, which have proved too demanding to manufacture (Offer and Bland–Hawthorn 1998). Holographic gratings have had more success (Blais–Ouellette 2004; Blais–Ouellette *et al.* 2004), but are still unable to achieve deep efficient suppression for a sufficient number of OH lines. High-dispersion masking has been adopted for several instruments (Iwamuro *et al.* 1994, 2001; Maihara and Iwamuro 2000; Maihara *et al.* 1993; Motohara *et al.* 2002). In this technique, the spectrum is first dispersed at high resolution, then the OH lines are rejected with a custom designed mirror-mask, before either re-dispersing at lower resolving power, or even recombining into white light. However, difficulty in maintaining the alignment of the mask, and also with scattering if not recombining into white light, has frustrated efforts in this method.

However, all of these requirements are able to be met using fibre Bragg gratings (FBGs), and indeed these have been demonstrated in on-sky experiments. We will first discuss the operation of FBGs, then their adaptation to the OH suppression problem. After this we will discuss their implementation in astronomical instruments, and show results from on-sky experiments. Thereafter, we will discuss future astrophotonic instruments using FBGs.

8.2.1 *FBGs for OH suppression*

The principle of Bragg gratings was introduced in Section 5.3. Assuming that a simple periodic FBG will suppress one OH line, then all other Bragg wavelengths must be outside the bandpass of interest. That is the free spectral range, $\delta\lambda$ must be larger than the bandpass.

Example 8.1: The astronomical H band is $\approx$300 nm wide, and has a central wavelength of 1.6 μm. For a FBG with a core index of 1.5, what must the grating period be to ensure that the FSR is wider than the bandpass? Rearranging Equation 5.40 gives

$$\Lambda = \frac{\lambda^2}{2n\delta\lambda},$$

and inserting the appropriate values yields

$$\Lambda = \frac{1.6^2}{2 \times 1.5 \times 300 \times 10^{-3}} \approx 2.8\ \mu\text{m}.$$

Example 8.2: For the same grating as in Example 8.1, how long would the grating need to be in order to achieve a resolving power of 10,000?

The resolving power is given by Equation 5.38. Rearranging, and inserting the appropriate values,

$$L = \Lambda R = 2.8 \times 10{,}000 = 2.8\ \text{cm}. \tag{8.1}$$

Example 8.3: For the same grating as in Examples 8.1 and 8.2 what is the required refractive index variation to achieve a suppression of 30 dB?
A suppression of 30 dB is equal to a reflectivity of $R = 1 - 0.001 = 0.999$.
From Equation 5.35 this yields a value of

$$\kappa = \frac{\tanh^{-1}\sqrt{R}}{L} = \frac{\tanh^{-1}\sqrt{0.999}}{2.8} = 1.48\ \mathrm{cm}^{-1}. \tag{8.2}$$

Then from Equation 5.36, the refractive index variation is

$$\Delta n = \frac{\kappa\lambda_{\mathrm{B}}}{\pi} = \frac{1.48 \times 1.6 \times 10^{-4}}{\pi} = 7.5 \times 10^{-5}. \tag{8.3}$$

Examples 8.1, 8.2, and 8.3 show that the requirements to achieve a single strongly suppressed OH line within and astronomical bandpass is feasible. However, in order to suppress more than one line a more complicated grating is needed. Naively, it is possible simply to splice together many single notch FBGs in series. However, for a splice with a throughput of α, and N lines, the total throughput would be

$$\eta_{\mathrm{total}} = \alpha^{N-1}, \tag{8.4}$$

which even for a 99% throughput at each splice yields a total throughput of only $\eta_{\mathrm{total}} = 37\%$ for $N = 100$. Thus, if many lines are to be suppressed, then splicing individual FBGs is to be avoided.

A more sophisticated method was found by Bland–Hawthorn *et al.* (2004), and later perfected by Bland–Hawthorn *et al.* (2008). This technique uses aperiodic variations in both the period and amplitude of the refractive index profile to suppress up to 150 lines in an individual FBG.

The power of this technique is best appreciated by examining the exquisite results which have been achieved. The top panel of Figure 8.5 shows the measured night sky spectrum using the GNOSIS instrument, which will be discussed further, and displays the familiar forest of OH emission lines. The middle panel shows the measured response of the GNOSIS FBGs. There are two FBGs in series, distinguished by the black and grey lines, which together suppress 103 OH doublets between 1.47 and 1.7 μm. The notches are perfectly matched to the OH lines in wavelength, and are also tailored to the lines in terms of the suppression factor; lines which are brighter on average have deeper notches; tailoring the notches in this way enables a greater number of notches to be written into a single FBG. The inset panels in the bottom panel show some details of the measured FBG transmission compared to a model OH spectrum. The

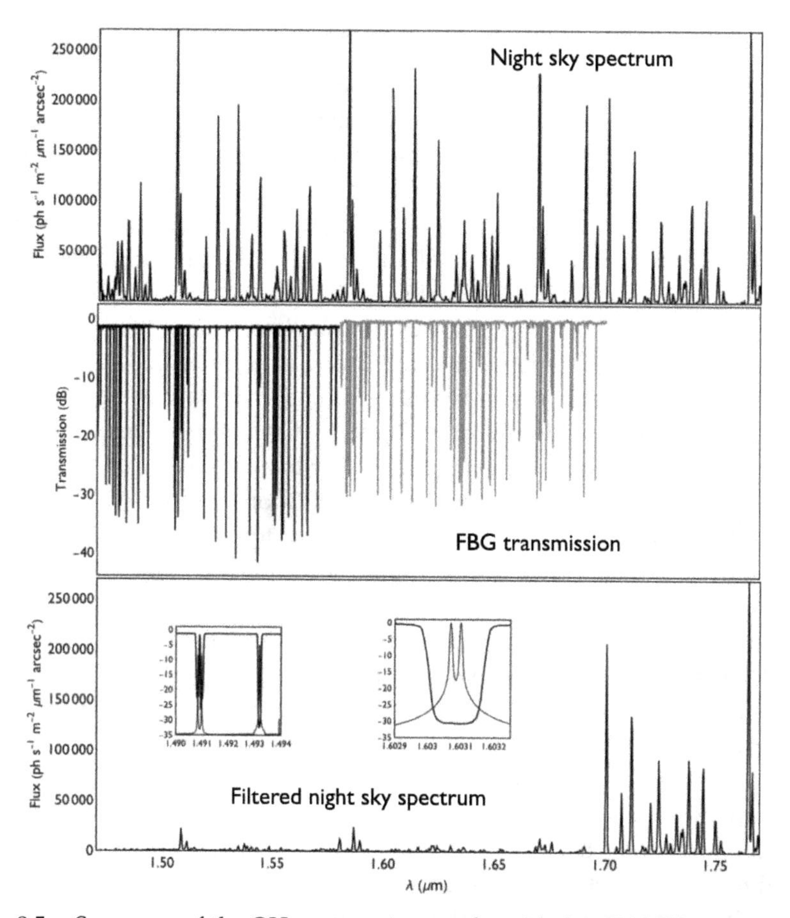

Fig. 8.5. Summary of the OH suppression results with the GNOSIS instrument. The top panel shows the night sky spectrum as measured with a control fibre containing no FBGs. The middle panel shows the measured transmission of two FBGs in series (black and grey). The notches are perfectly tuned to the OH lines in wavelength and in depth. The inset panels in the bottom panel illustrates the narrowness and squareness of the notches. The bottom panel shows the night sky spectrum after OH suppression. The lines between 1.47 and 1.7 μm are all strongly suppressed, and the integrated background is reduced by a factor of 9.

notches are both deep, with up to 40 dB of suppression, and very narrow, at $\Delta\lambda \approx 200$ pm, and very square. The squareness is desirable since it ensures that only the OH lines are suppressed, and not the surrounding spectrum. The notches can be made deliberately wider if the doublet spacing is larger.

8.2.2 *Implementing FBGs in an astronomical instrument*

It has been seen in the previous discussion that aperiodic FBGs are capable of producing extremely complicated notch filters which can be almost perfectly matched to the OH lines. This remarkable performance is an excellent demonstration of the power and potential of astrophotonics. We now discuss how FBGs can be incorporated into an astronomical instrument.

FBGs only work in single-mode fibre. This is because each mode of a multimode fibre has a different effective index, and therefore for a particular grating period the Bragg wavelength will be different for each mode, see Equation 5.33. Writing an Bragg grating in a multimode fibre would result in a notch that is smeared out in a broad and shallow dip due to the different Bragg wavelengths of each mode. On the other hand, we have seen that efficient coupling from a telescope to a fibre requires multimode fibre in all but perfect diffraction limited conditions (Chapter 6). Therefore, to use FBGs we must convert multimode fibre to an array of single-mode fibres. This is exactly what is achieved with photonic lanterns (Section 6.3.2), and indeed photonic lanterns were first developed for this purpose, although they have since found many non-astronomical applications.

Thus, we are able to develop a scheme for incorporating FBGs into a fibre feed for a spectrograph. A telescope feeds a multimode fibre, which is converted into an array a single-mode fibres via a photonic lantern. To each single-mode fibre is spliced an identical aperiodic FBG. These are then spliced into a reverse photonic lantern, which converts the individual single-mode fibres back to a multimode fibre. The multimode fibre is then used to feed a spectrograph, see Figure 8.6.

Using this basic scheme the GNOSIS instrument (Trinh *et al.* 2013b) was used to demonstrate OH suppression with FBGs by using a combination of seven photonic lanterns each feeding 2×19 FBGs to feed an existing instrument IRIS2 (Tinney *et al.* 2004) on the Anglo-Australian Telescope. The GNOSIS design has been improved upon with its successor, PRAXIS, which uses a vacuum feed-through such that the fibre output slit is contained within the spectrograph dewar, which along with cooling of the fore-optics, minimises the thermal emission from the OH suppression unit (Ellis *et al.* 2020).

The results of these experiments convincingly demonstrated the performance and potential of FBGs for OH suppression (see Ellis *et al.* 2012a, 2020; Trinh *et al.* 2013a,b for full details). Figure 8.5 shows the night sky background measured with GNOSIS. The FBGs suppress 103 OH doublets

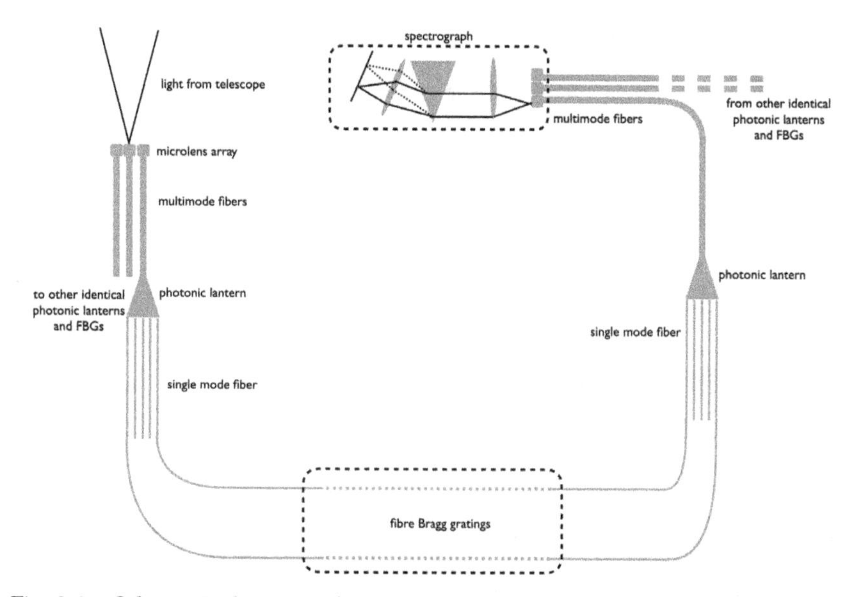

Fig. 8.6. Schematic drawing of an instrument employing FBGs for OH suppression. Both the GNOSIS and PRAXIS instruments were based on this concept.

between 1.47 and 1.7 μm, and the integrated background over these wavelengths was reduced by a factor of 9.

The PRAXIS instrument demonstrated OH suppression in a high efficiency (18%) spectrograph for the first time (Ellis *et al.* 2020). Figure 8.7 which shows PRAXIS spectra of the Seyfert galaxy NGC 7674, for various degrees of sky-subtraction and OH suppression. The top line shows the raw spectrum with no OH suppression, nor sky-subtraction; the middle line shows the spectrum after OH suppression, but with no sky-subtraction, and the bottom line shows the OH suppressed sky-subtracted spectrum, revealing emission lines hidden in the noise of the initial raw spectrum. The remarkable difference in starting points between the non-suppressed purple spectrum and the suppressed blue spectrum shows the advantage conferred by OH suppression.

Unfortunately PRAXIS still suffered from high thermal emission, and the full potential of OH suppression has still not been realised at the time of writing, although a successor to PRAXIS, named Eupraxia, is being planned for use on an 8 m class telescope. However, it should be noted that this thermal emission is *not* due to the photonic components; indeed the astrophotonic components of PRAXIS all perform extremely well and reliably.

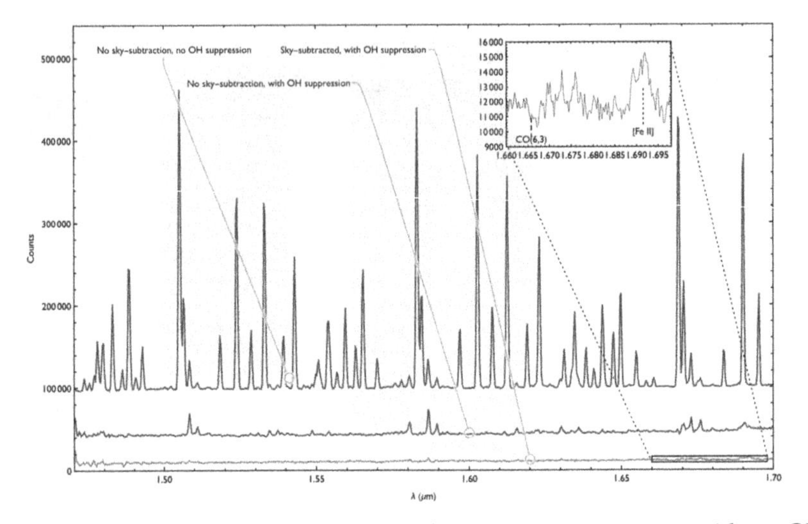

Fig. 8.7. PRAXIS spectra of NGC 7674, showing a raw spectrum with no OH suppression nor sky-subtraction, a raw spectrum with OH suppression but no sky-subtraction, and a sky-subtracted, OH suppressed spectrum. Reproduced with permission from Ellis *et al.* (2020). © Oxford University Press.

8.3 Current and Future Developments in Bragg Gratings for OH Suppression

The GNOSIS and PRAXIS experiments have shown that the combination of photonic lanterns and FBGs can significantly reduce the sky-background for near-infrared spectroscopy in a high efficiency instrument. Nevertheless, there is potential for improvement, not just in terms of implementation (e.g. PRAXIS vs. GNOSIS), but also in terms of the photonic components.

In particular, the combination of conventional photonic lanterns and FBGs is an inefficient method to scale-up to larger instruments. PRAXIS covered a single seeing-limited source of 1.65 arcsec with seven fibres, each of which fed 19×2 FBGs, making 266 FBGs in total. Each of these FBGs had to be written individually, and then spliced into the photonic lanterns, requiring 399 separate splices in total.

Here, we discuss some other techniques currently under development aimed at improving OH suppression. We confine our discussion to a brief overview of each technique, and a summary of current developments, and refer the reader to the references for full details.

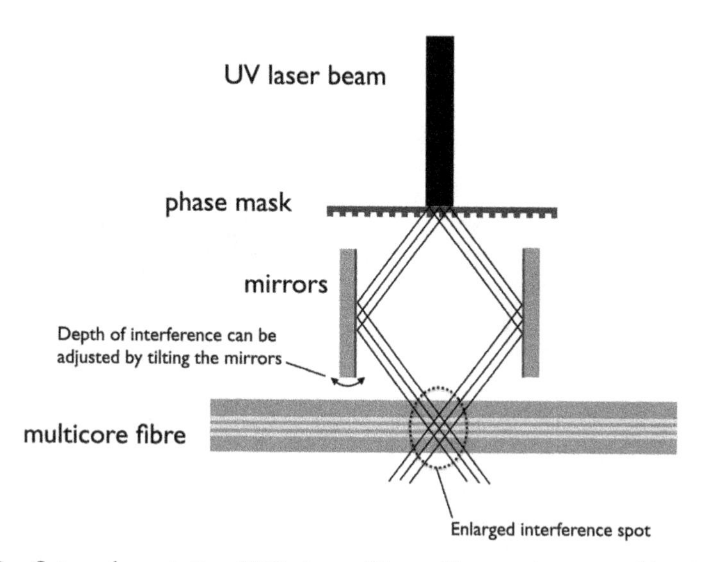

Fig. 8.8. Set-up for printing FBGs in multicore fibres, using a combination of a phase mask and a Talbot interferometer, following Bland–Hawthorn *et al.* (2016).

8.3.1 *Multicore fibre Bragg gratings*

The efficiency of inscribing FBGs can be increased using multicore fibre photonic lanterns (Section 6.3.2.1). In this case, the individual cores of the multicore fibre can all be inscribed with the same Bragg grating simultaneously, both speeding up the process of writing the gratings, and significantly reducing the size of the final device, which is also more easily handled.

In order for this technique to work several subtle issues must be resolved. Writing FBGs is usually done by using a UV laser to create an interference pattern along the length of the fibre, either by using a phase mask to create the interference pattern, or via an interferometer. This interference pattern breaks the molecular bonds in the fibre core, creating a periodic change in refractive index. For a multicore fibre the interference pattern must cover all cores of the fibre simultaneously, and this requires a larger spot of the interference pattern. This can be achieved using a Talbot type interferometer combined with a phase mask (Bland–Hawthorn *et al.* 2016) as in Figure 8.8.

Second, the intensity across all cores must be equal, otherwise the strength of the induced refractive index variations will be different, leading to a shift in Bragg wavelength and reflectivity, see Equations 5.33

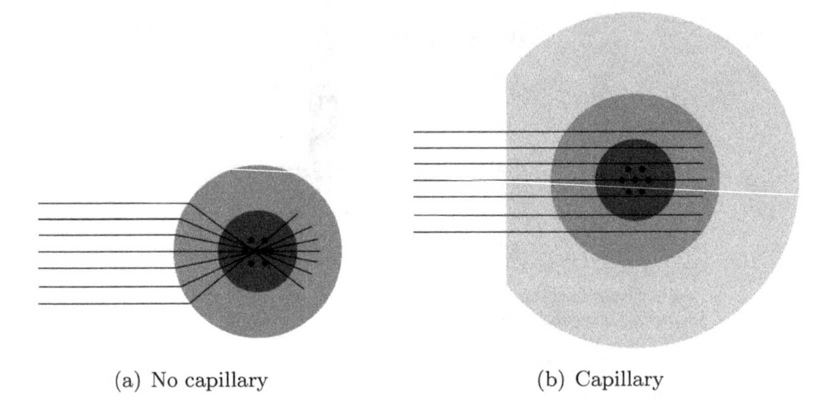

(a) No capillary (b) Capillary

Fig. 8.9. (a) The lensing effect due to the cladding when writing a Bragg grating in a multicore fibre cause a variation in the intensity of the laser over the core of the fibre. In extreme cases, the outer cores could be missed altogether. (b) If the fibre is placed inside a capillary tube which has one side polished flat, the lensing effect can be circumscribed.

and 5.35. This is difficult as the cladding of the fibre acts as a lens, see Figure 8.9, but has been solved by placing the multicore fibre (MCF) inside a capillary tube with one side polished flat (Lindley 2017; Lindley *et al.* 2014).

Even when using a capillary there is a consistent offset in the Bragg wavelengths of a multicore fibre as a function of radius (Lindley *et al.* 2014). This is manifest in Figure 8.10 in the difference between the central and the outer cores. This issue is not yet fully understood, but seems to be a property of MCF itself (Mosley *et al.* 2014), and not due to the FBG writing process. It may be that the cross-coupling of the electric field between cores (Section 6.5.2) is responsible for the shift in effective index. For example the central core of a 7 core MCF has six nearest neighbours, whereas the outer cores only have three nearest neighbours. This difference could cause the shift in effective index. Note that this shift is only of the order $\Delta n \sim 10^{-5}$. Nevertheless, the fabrication of MCF is good enough to ensure that differences in core radius or core shape cannot be responsible for the shifts (Ellis *et al.* 2018).

8.3.2 *Direct-write waveguide Bragg gratings*

Bragg gratings can also be written into other waveguides besides fibres. One interesting alternative is to inscribe Bragg gratings into direct-write

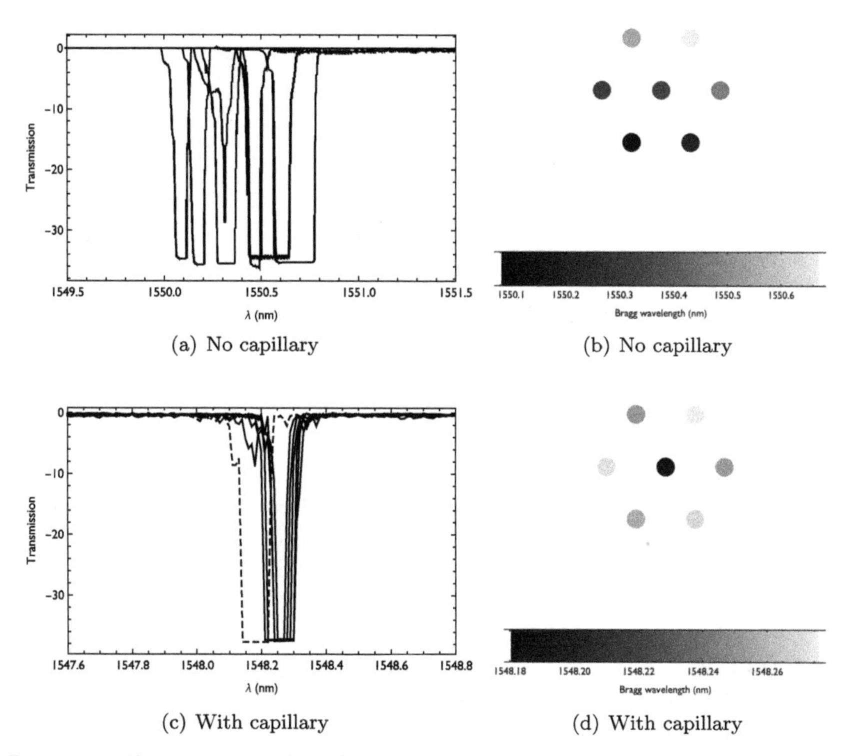

(a) No capillary

(b) No capillary

(c) With capillary

(d) With capillary

Fig. 8.10. The Bragg wavelengths in a 7 core MCF, written without any compensation for the lensing effect (top panels), and using a polished capillary to minimise the lensing effect (bottom panels). The FBGs written with the capillary are much more uniform, with the exception of the central core.

waveguides (Section 5.5). The Bragg gratings can be written either point-by-point after the waveguide itself is written (e.g. Marshall *et al.* 2006) or as part of the same process in which the waveguide is written (e.g. Zhang *et al.* 2007).

The attraction of direct-write Bragg ratings is not simply as an alternative to FBGs, but rather because they are more easily integrated with photonic lanterns, since both can be written at the same time as part of the same process (Spaleniak 2014; Thomson *et al.* 2009). Indeed, Spaleniak (2014) has explored the application of direct-write Bragg gratings and photonic lanterns for astronomy, see Figure 8.11.

Figure 8.12 shows the inscription of four Bragg gratings in the same waveguide, in which defects are written point-by-point in the waveguide, and the gratings are written side-by-side. Even though the defects do not

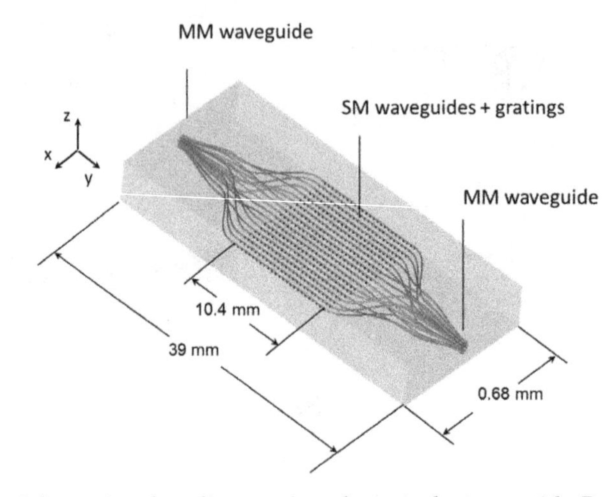

Fig. 8.11. Schematic of a direct-write photonic lantern with Bragg gratings inscribed into the single-mode waveguides. Reproduced with permission from Spaleniak (2014).

fill the entire area of the waveguide core, the defects still act to modify the effective index at that point, causing a reflection as for a FBG.

Photonic lanterns written in this way have achieved end-to-end throughputs of 60–75% supporting up to 19 single-mode waveguides, very similar to the performance of fibre based lanterns (Spaleniak *et al.* 2014; Thomson *et al.* 2011). One difference is that direct-write waveguides have smaller NA, and must be fed with correspondingly slower beams of $\sim f/8$ or so. Analogously to fibres, multimode waveguides will suffer from FRD if fed at much slower beams ($>f/10$), but if fed close to the NA of the waveguide FRD is minimal (Jovanovic *et al.* 2012a).

Prototypes chips have been made which can suppress up to four separate lines by a factor of $\approx$5 dB (Spaleniak *et al.* 2014). Longer gratings should achieve deeper notches. Although, the point-by-point inscription facilitates writing multiples notches, it seems that this method will be limited to devices suppressing only a few lines at a time. Therefore, to cover a wider wavelength range the light from the telescope must first be divided into separate wavelength channels, each of which feeds a separate device, as sketched in Figure 8.13, from Spaleniak (2014). Alternatively, aperiodic Bragg gratings, such as the FBGs used in GNOSIS, could be developed by varying the refractive index of the waveguide as it is written. In principle,

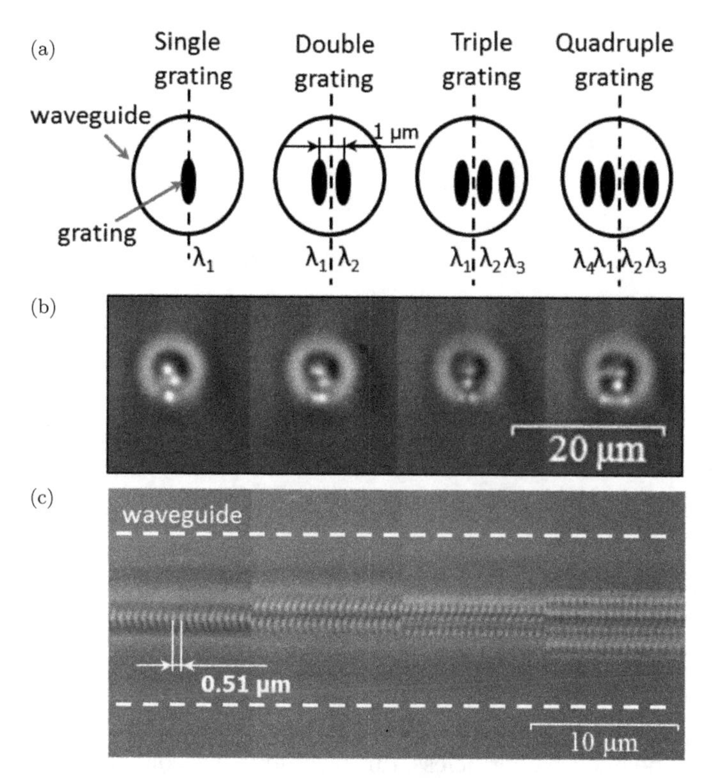

Fig. 8.12. The inscription of a multi-notch Bragg grating in a direct-write waveguide. The waveguide itself is written first, and thereafter each grating is written in a point-by-point manner in turn. (a) A sketch of the process; each grating has a different period, and the gratings are written side-by-side in the waveguide core. (b) A photograph of the end-face of a waveguide with four gratings, at four different locations along the waveguide. (c) A top-down view of waveguides with 1, 2, 3, and 4 gratings. Reproduced with permission from Spaleniak (2014).

this could achieve suppression of a large number of wavelengths simultaneously, but to our knowledge no work has been done in this direction, and thus would require significant development.

8.3.3 *Direct-write into fibres*

The direct-write inscription of Bragg gratings can also be applied to fibres. Goebel *et al.* (2018) have demonstrated a 10 notch FBG covering an 80 nm bandpass from 1510–1590 nm in this way. These gratings were not written

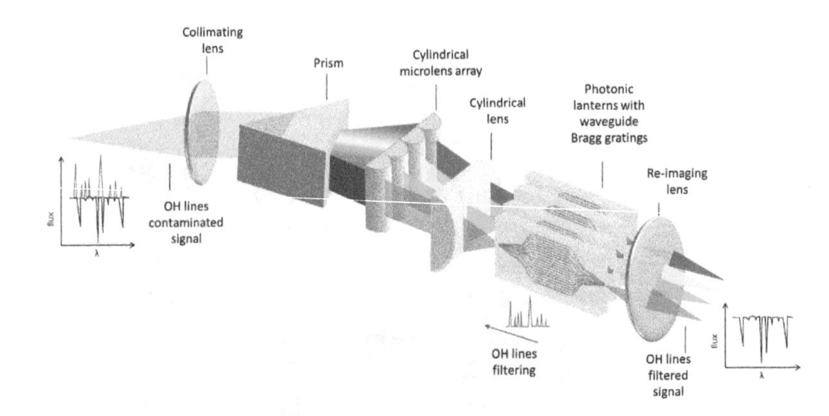

Fig. 8.13. Schematic showing how direct-write OH suppression chips can be incorporated into an astronomical instrument to suppress a large number of OH lines. Reproduced with permission from Spaleniak (2014).

with overlapping individual point-by-point gratings, but rather used an aperiodic grating as for the FBGs described in Section 8.2.1. So far FBGs written in this way have been physically short at ≈50 mm, and the notches have been correspondingly shallow (~4 dB). Using a higher power laser can increase the depth to ~25 dB, but at the expense of large losses (~5 dB) between the lines. Nevertheless, this technique has considerable promise, especially as it can be easily adapted for non-standard fibre such as fluoride (Fuerbach *et al.* 2019; Bharathan *et al.* 2019).

8.3.4 *Channel waveguide Bragg gratings*

Another type of waveguide Bragg grating currently being developed for OH suppression uses lithographic channel waveguides (Section 5.4). In this case, the refractive index variation is provided by slightly altering the width of the waveguide along its length, see Figure 8.14.

This method shows significant promise, with early prototypes already achieving 20 notch devices matched to within 100 pm of the target wavelengths, with suppression depths of 15–33 dB and a bandwidth of 300–400 pm, see Figure 8.15 (Zhu *et al.* 2016).

So far channel waveguide Bragg gratings have not been incorporated into photonic lanterns or astronomical instruments. As discussed in Section 6.4.4 this requires an extra-step in the integration of the waveguides into an instrument, since after the photonic lantern coupling to SMFs

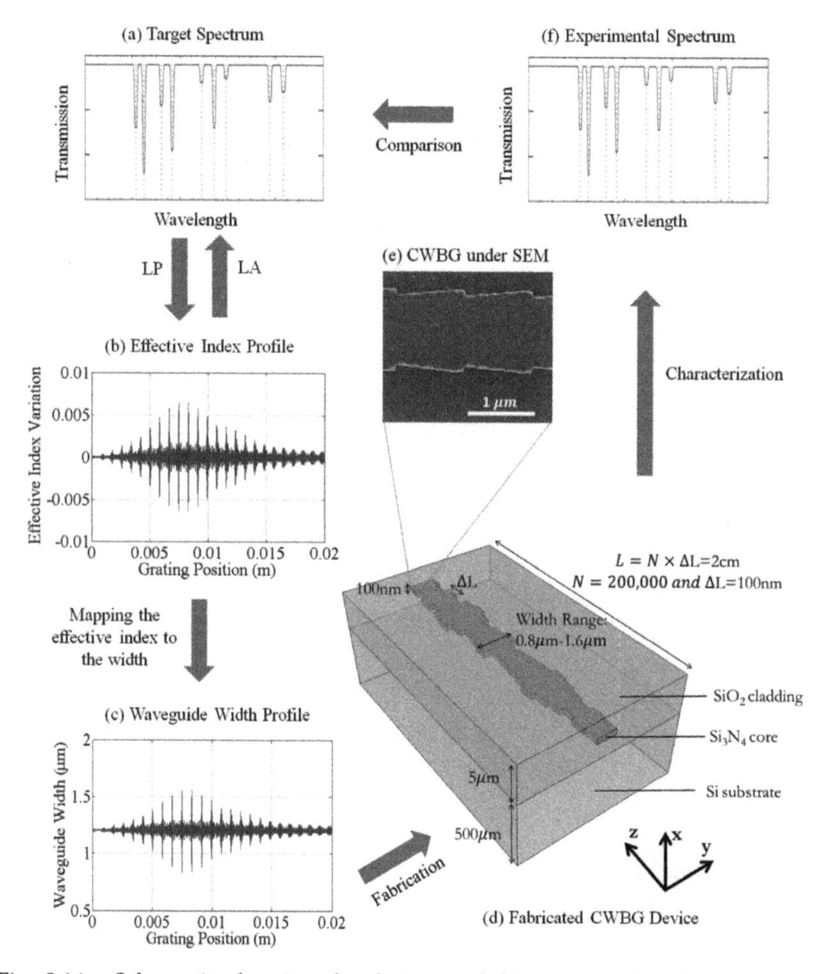

Fig. 8.14. Schematic showing the design and fabrication of a channel waveguide Bragg grating. Reproduced with permission from Zhu *et al.* (2016). © AIP Publishing.

there is the extra challenge in coupling into the channel waveguides. Moreover, this must be achieved in a 'packaged' solution, i.e. one that is fixed and does not require alignment or adjustment, and one which can also accept multiple fibre inputs. Once this packaging issue is solved channel waveguide Bragg gratings could become very competitive since the lithographic process of production is highly reproducible and modular; see the discussion in Chapter 9.

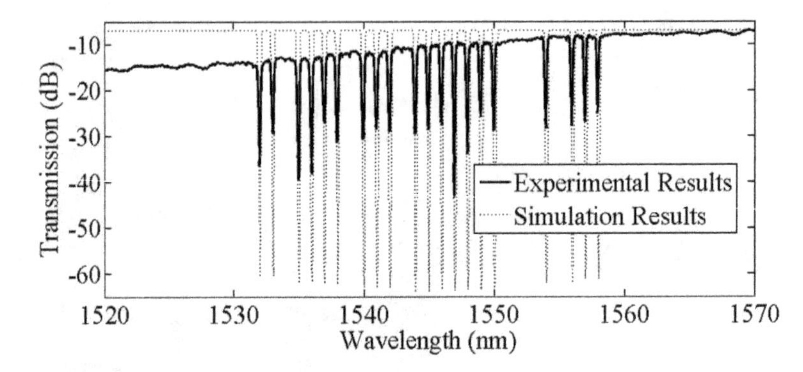

Fig. 8.15. Laboratory results for a channel waveguide Bragg grating showing 20 notches matched to the target wavelengths with 15–30 dB of suppression. Reproduced with permission from Zhu *et al.* (2016). © AIP Publishing.

8.4 Ring Resonators

A somewhat different solution to OH suppression, which does not rely on Bragg gratings, is to use ring resonators. These are based on channel waveguides, as shown in Figure 8.16. Light from the input waveguide evanescently couples into the ring, whereupon it constructively and destructively interferes with itself until only light at the resonant wavelengths of the cavity remains. The condition for resonance is therefore

$$m\lambda = nL, \tag{8.5}$$

where m is an integer, λ is the wavelength, n is the effective index and L is the circumference of the ring. The resonant light couples back into the input waveguide and interferes with the input light. At each coupling from one waveguide to another there is a $\pi/2$ phase change (Section 6.5.2), thus the light coupled back to the input waveguide is now π radians out of phase with the input light, and therefore the interference is destructive and effectively filters the input signal at the resonant frequencies of the ring. Thus, a series of ring-resonators, each tuned to the wavelength of a different OH night sky line, could provide a means of OH suppression, see Figure 8.17.

The potential application of ring resonators for OH suppression has been examined in some detail by Ellis *et al.* (2017). We illustrate the performance and requirements of ring resonators for astronomical instruments with the simple arrangement shown in Figure 8.18, in which a single bus waveguide is coupled to a single ring. The behaviour of more complex

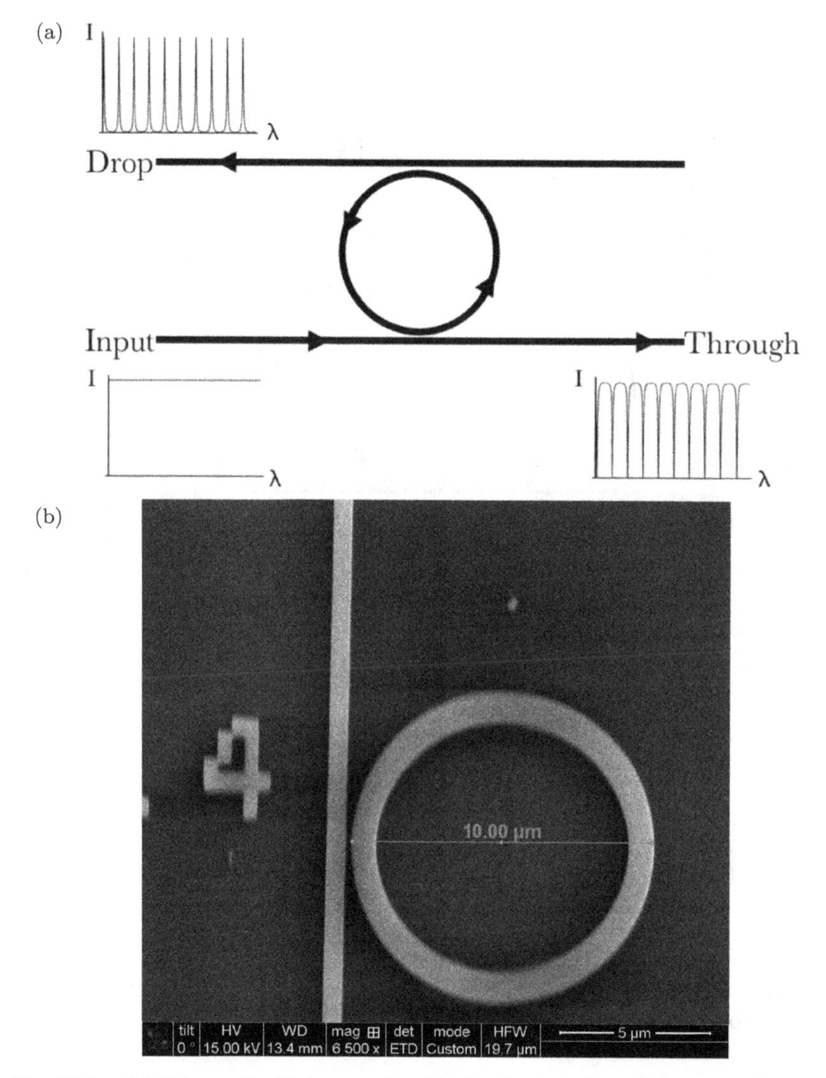

Fig. 8.16. (a) Schematic diagram of a simple ring resonator showing the input, through and drop ports and a sketch of the spectrum at each port (b) SEM photograph of a Si_3N_4 ring resonator with a through port and no drop port.

devices can be derived in a similar manner. However, in order to avoid excessive derivations we omit these, and give only the most pertinent results.

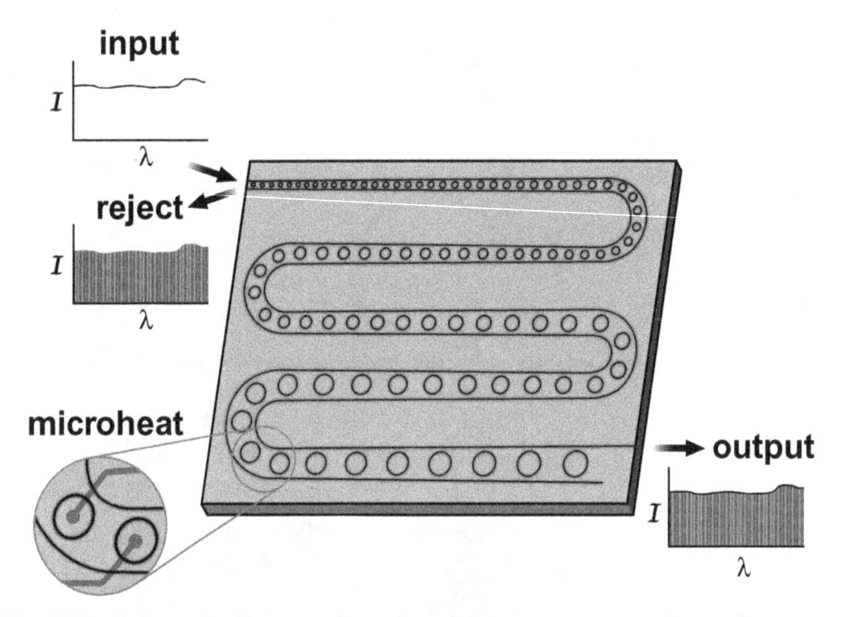

Fig. 8.17. Schematic diagram of a series of ring resonators, each tuned to a particular OH line, and stabilised with a microheater, acting as a multi-notch OH suppression unit.

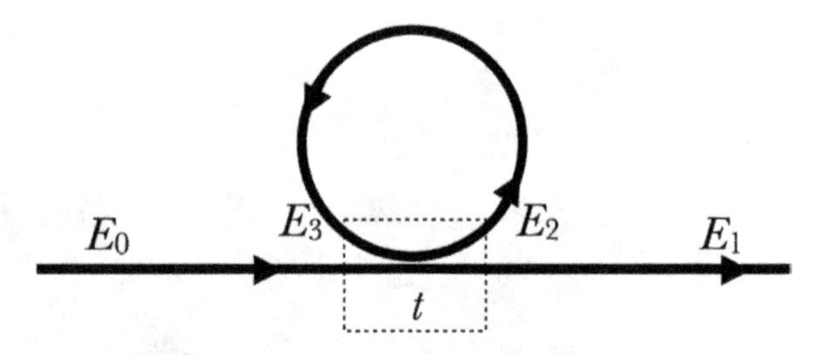

Fig. 8.18. Schematic diagram of the electric fields and coupling coefficients in a simple resonator with no drop port.

Let the throughput of one passage around the ring be α, and the phase change of one passage around the ring be θ. The fraction of the field that couples into the second waveguide depends upon their separation, length, and the refractive index profile of the media. Consider the schematic coupling depicted in Figure 8.19, where the self-coupling coefficient is t and the cross-coupling coefficient is $\kappa e^{-i\frac{\pi}{2}} = -i\kappa$, which takes into account the $\pi/2$ phase shift which occurs when light is evanescently coupled between waveguides.

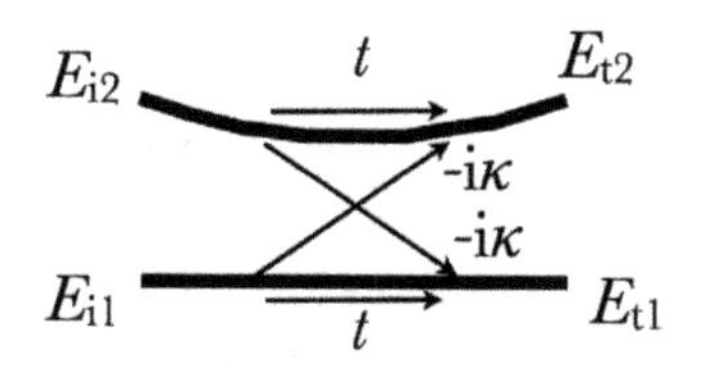

Fig. 8.19. Schematic diagram of cross-coupling of the electric fields between two waveguides, and the notation used herein.

If we assume that the coupling is lossless then

$$t^2 + \kappa^2 = 1, \tag{8.6}$$

and we can express the relationship between the input electric fields E_{i1}, E_{i2} and the output fields, E_{t1}, E_{t2}, by the matrix relation

$$\begin{pmatrix} E_{t1} \\ E_{t2} \end{pmatrix} = \begin{pmatrix} t & -i\sqrt{1-t^2} \\ -i\sqrt{1-t^2} & t \end{pmatrix} \begin{pmatrix} E_{i1} \\ E_{i2} \end{pmatrix}. \tag{8.7}$$

Now, from after one passage around the ring the amplitude of the electric field has been diminished due to the loss, such that

$$E_3 = E_2 \alpha e^{-i\theta}. \tag{8.8}$$

Substituting this into Equation 8.7 eliminating E_2 and solving for E_1 yields

$$E_1 = E_0 \frac{t e^{i\theta} - \alpha}{e^{i\theta} - t\alpha}. \tag{8.9}$$

Therefore, the fractional intensity at the through port is given by

$$\left| \frac{E_1}{E_0} \right|^2 = \frac{t^2 + \alpha^2 - 2t\alpha \cos\theta}{1 + t^2\alpha^2 - 2t\alpha\cos\theta}. \tag{8.10}$$

The total phase change for a single passage around the ring may be expressed in terms of wavelength as

$$\theta = \frac{2\pi n L}{\lambda}, \tag{8.11}$$

where L is the circumference of the ring, n is the effective index of the waveguide and λ is the wavelength of the light.

8.4.1 *Requirements for OH suppression*

Equation 8.10 can be used to determine the requirements on ring-resonators for OH suppression. We give a synopsis of these requirements here; for further elaboration, see Ellis *et al.* (2017).

8.4.1.1 *Suppression factor*

The suppression should ideally reach ≈30 dB for the brightest lines; with diminishing returns for deeper notches, although notches of ≈20 dB will still be of value. Note from Equation 8.10 that the resonances occur when $\theta = 0$, and thus when $t = \alpha$ then the transmitted flux at the through port is zero, i.e. the suppression would be perfect under these conditions. Therefore, in order to achieve high suppression factors we require $t \approx \alpha$.

8.4.1.2 *Q factor or resolving power*

The OH lines are intrinsically very narrow, with a typical separation between the Λ-doublets of ≈100 pm (see the discussion at the start of this chapter), and thus, the bandwidth of the notches should also be very narrow, ≈200 pm, for optimal suppression whilst allowing some latitude for misalignment. The Q factor, or in astronomical parlance, the resolving power can be derived from Equation 8.10 by finding the wavelengths when the depth is equal to half the maximum, to give

$$\frac{\lambda}{\delta\lambda} = \frac{\pi n_g L}{\lambda} \frac{1}{\cos^{-1}\left(\frac{2t\alpha}{1+t^2\alpha^2}\right)}, \tag{8.12}$$

which is plotted in Figure 8.20 for ring with radius of 5 μm at a wavelength of 1.6 μm for different group indices of $n_g = 1.5, 2,$ and 3.4, as a function of $\alpha = t$. It can be seen from Equation 8.12 that the Q factor increases linearly with ring radius and group index, and from the figure that the Q factor increases very rapidly with t or α. Thus, as for the suppression depth it is necessary that both α and t are close to 1 and nearly equal.

8.4.1.3 *Free spectral range*

The free spectral range is one of the most demanding requirements on using ring resonators for OH suppression, since the astronomical passbands of interest are generally very large, e.g. the H band covers ~300 nm. A single ring will give a series of resonances corresponding to different integer values of m in Equation 8.5, or equivalently when $\theta = m2\pi$ is Equation 8.10, but these will be equally spaced in frequency whereas the OH lines are not. Thus, we can only align one notch to one OH line (or at best 2), and the other notches must be outside the wavelength region of interest or else they will filter light from the parts of the spectrum which are desired to be kept. In other words, the free spectral range must be larger than the wavelength range. The free spectral range, $\Delta\lambda$ follows

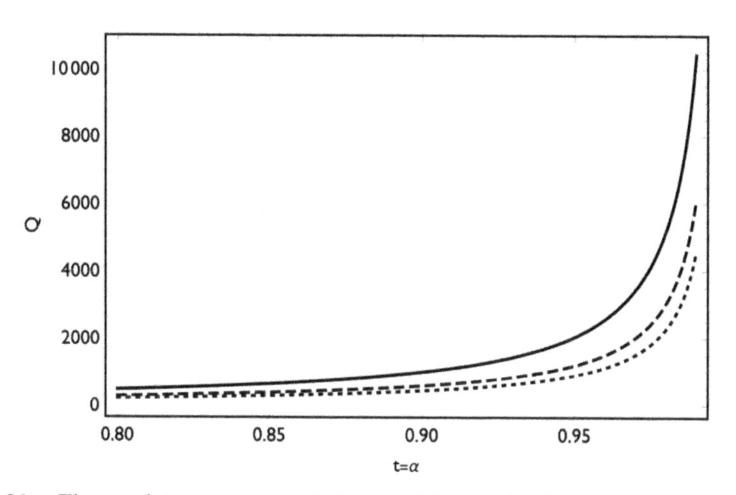

Fig. 8.20. The resolving power, or Q factor, of the notch of a simple ring resonator with radius = 5 μm at a wavelength of 1.6 μm for group index of 1.5 (dotted), 2 (dashed), or 3.4 (black), as a function of t, with $t = \alpha$.

directly from the differentiation of the resonance condition, Equation 8.5, yielding

$$\Delta\lambda = \frac{\lambda^2}{Ln_g}, \tag{8.13}$$

where

$$n_g = n - \lambda \frac{\partial n}{\partial \lambda}, \tag{8.14}$$

is the group index, which is equal to c/v_g, where v_g is the group velocity, which describes the velocity at which a pulse propagates along a waveguide (see Section 4.2.3). Note that if the refractive index is independent of wavelength $n_g = n$.

Figure 8.21 shows the FSR as a function of ring radius for devices with group index $n_g = 1.5$, 2, and 3.4, at a wavelength of 1.6 μm. These results present something of a dilemma for ring resonator design. It is clear that a large FSR requires a small radius. However, a small radius will result in unacceptable bending losses (low α) unless a high refractive index is used to confine the waveguide mode. Unfortunately, a high refractive index results in decreasing the FSR, and therefore partly counteracting the effect we are trying to achieve.

Since, a high Q also requires a high refractive index, it seems that ring resonators will require the passband to be divided into smaller wavelength

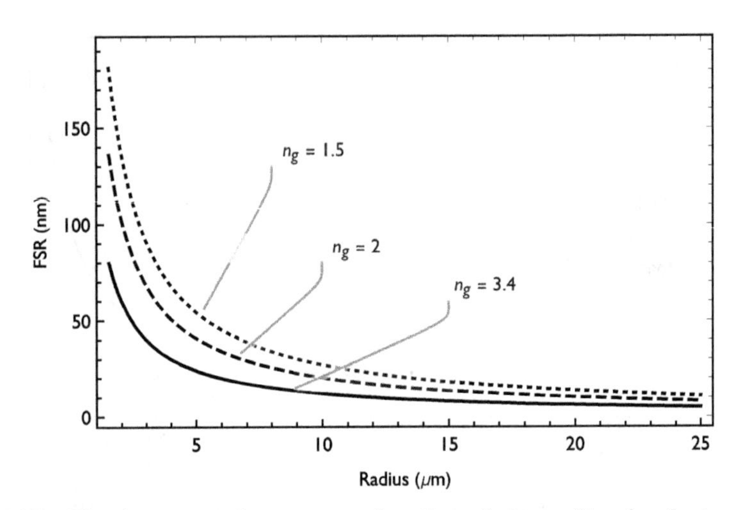

Fig. 8.21. The free spectral range as a function of ring radius for devices with group index $n_g = 1.5$, 2, and 3.4, at a wavelength of 1.6 μm.

ranges before the light is fed to the ring resonators, as for direct-write FBGs, see Figure 8.22, cf. Figure 8.13.

A possible solution to increasing the FSR without using very small radius rings is to use Vernier coupled rings as sketched in Figure 8.23. Each ring has a different circumference, but both are designed to be resonant at a particular frequency. So, we have

$$m\lambda = nL_1,$$
$$p\lambda = nL_2.$$

And if $m = M$ and $p = P$ for the resonant wavelength λ_0 which is common to both rings, and M and P are chosen to be coprime (i.e. they have no common divisors), then every Mth notch of ring 1 will overlap with every Pth notch of ring 2, with no overlapping in between. In this way, it is possible to increase the free spectral range of the resonator without requiring very small resonators as implied by Equation 8.13, which will eventually become limited by bending losses for small radii. Thus, we have

$$\frac{M}{P} = \frac{L_1}{L_2}, \tag{8.15}$$

so the ratio of the circumferences of each ring should be a rational number.

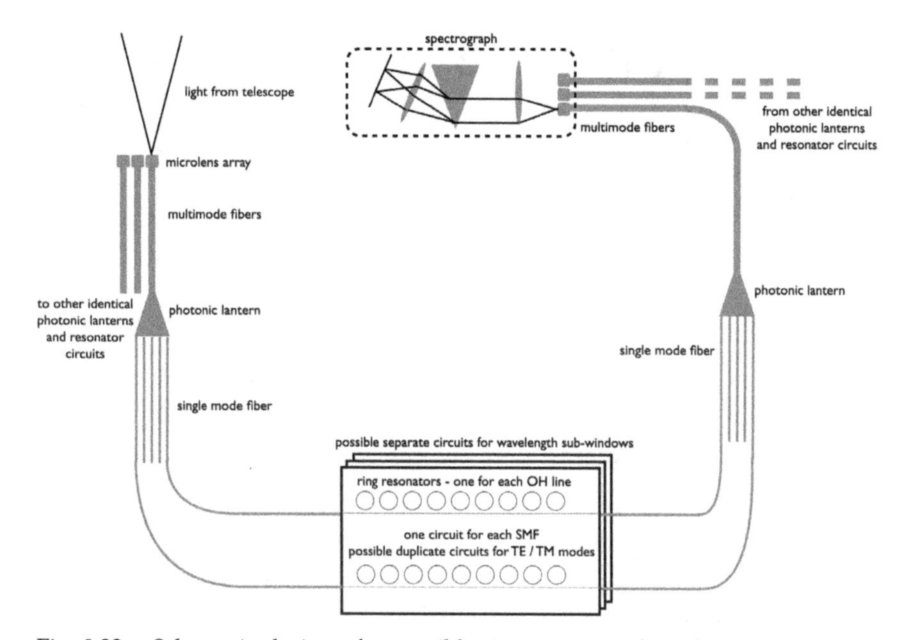

Fig. 8.22. Schematic design of a possible ring resonator based OH suppression system.

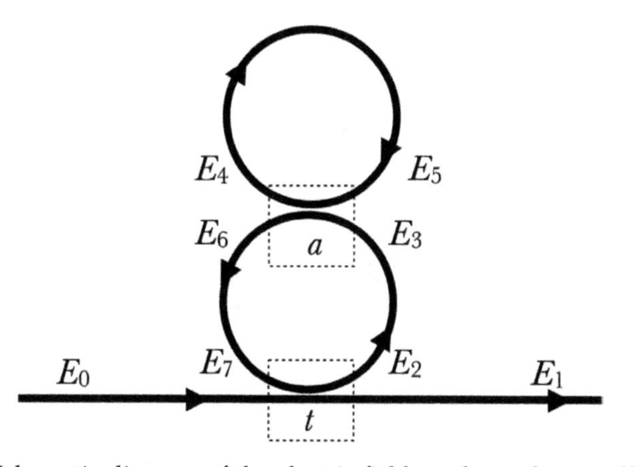

Fig. 8.23. Schematic diagram of the electric fields and coupling coefficients in a Vernier coupled resonator with no drop port. The transmission at the through port can be solved in the same way as before.

Figure 8.24 shows an example transmission function with $\lambda_0 = 1.6$, $M = 9$ and $P = 10$, compared to the transmission function of devices made from the individual rings.

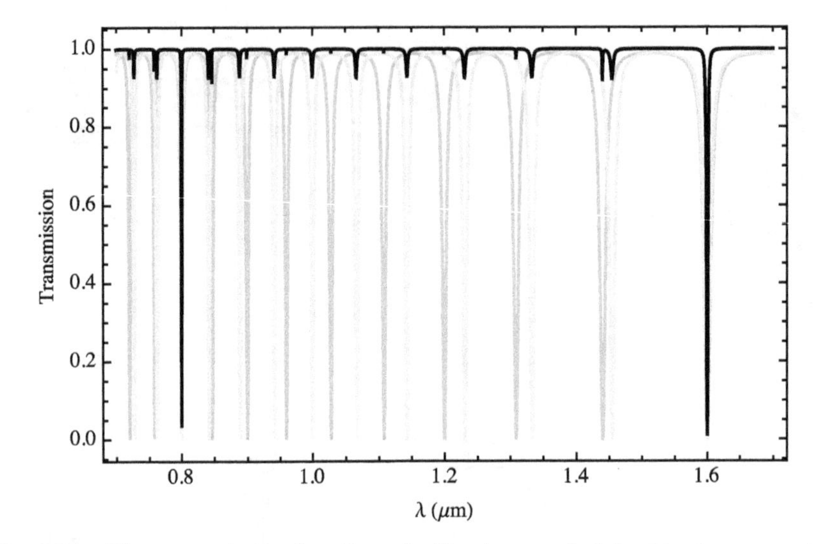

Fig. 8.24. The transmission function of a Vernier coupled double ring resonator (black) with $R_1 = (9/10)R_2$, and $\alpha = 0.998$, $\beta = 0.98$, and $t = 0.9$, compared to devices made from the individual rings (red and blue) with $\alpha = t = 0.9$.

Prima facie, Vernier rings are a promising solution to realising large FSR ring resonators. However, there are tight requirements on matching the coupling coefficients and throughputs between the rings which may be difficult to achieve in practice, and laboratory demonstrations are necessary in order to prove the method.

8.4.1.4 *Polarisation*

Because channel waveguides are rectangular in cross-section, they will generally support both TE and TM modes, and each mode will have a different effective index (Section 4.2.3). Therefore, a ring resonator may actually have a double set of resonances, slightly offset from one another in wavelength, one for each polarisation.

There are three possible solutions to this problem. First, one could simply filter out one polarisation, either before the light enters the waveguide, or else by using a waveguide which only supports one mode, e.g. a very oblong waveguide. However, since most astronomical sources are unpolarised this will result in a loss of $\approx 50\%$ immediately, which is not a very satisfactory outcome.

Second, one could split the incoming light into two orthogonal polarisations before it enters the waveguides. In this case, the number of circuits would need to be doubled, with duplicates for each polarisation. Since there may already be multiple circuits, one for each spatial mode of the multimode fibre, and one for each wavelength subregion of the passband, this could result in a large total number of circuits, as in Figure 8.22. However, since it is possible to fit many hundreds of circuits on a single wafer, this is not necessarily a serious impediment; indeed the small size, modularity and replicability is one of the major attractions of using ring resonators and other lithographic waveguides.

Third, it is possible to make polarisation independent waveguides, through careful control of the waveguide geometry. However, since the mode profiles, and therefore the effective index of a waveguide is dependent on wavelength, this will not work over large wavelength ranges. More work is required to see if polarisation independent waveguides can be made to work over the FSR required.

8.4.1.5 *Laboratory tests*

Ring resonators have not yet been incorporated into photonic lanterns or astronomical instruments. Nevertheless, laboratory tests show promise with SOI wafers. Current devices can now achieve the required depth, FSR $\approx$30 nm, and Q factor simultaneously, and multiple wavelengths and tuning of the wavelengths has also been demonstrated separately (Kuhlmann *et al.* 2018; Liu *et al.* 2021).

The major limiting factor is currently the overall throughput. The major losses associated with ring resonators are the insertion losses, bend losses and absorption, and of these the insertion losses are by far the hardest to reduce, see the discussion in Section 6.4.4. However, it should be noted that the requirements for astronomical instruments are somewhat different than for telecomms. in that a bespoke solution is permissible for an astronomical instrument since the devices will not required mass production, but rather will be used for individual instruments. Thus, a solution which requires more input in the fabrication may be acceptable. On the other hand, the required throughput will usually be higher for astronomical devices than for telecomms., since the sources are intrinsically faint.

CHAPTER 9

PHOTONIC SPECTROGRAPHS

Consider a telescope observing a star of angular diameter, Γ arcsec, determined by the seeing (Section 2.1). The size of the spot formed at the focal plane will be

$$\text{spot size} = \frac{\Gamma f D}{206{,}265}, \tag{9.1}$$

where f is the telescope focal ratio and D is the telescope mirror diameter (Section 2.2.4). Therefore, for a specific focal ratio, as telescopes get larger, so does the spot size. This in turn leads to larger instruments with larger optics. For the next generation of extremely large telescopes (ELTs) with diameters over 20 m the instruments are very large indeed, see for example Figure 9.1. Of course, larger instruments are also more expensive and more demanding to make.

This scaling relation can be broken for fibre fed instruments by using microlens arrays and photonic lanterns to divide the beam into single-mode fibres. Spectrographs for single-mode fibres are by definition diffraction limited. Diffraction limited spectrographs can be made very small. Thus, the problem is changed from one of producing extremely large spectrographs, to one of replicating very many extremely small spectrographs. This type of modularity and reproducibility is one of the key strengths of photonics, especially for lithographic photonic circuits. Thus, as the next generation of ELTs approaches the case for astrophotonic spectrographs becomes more compelling.

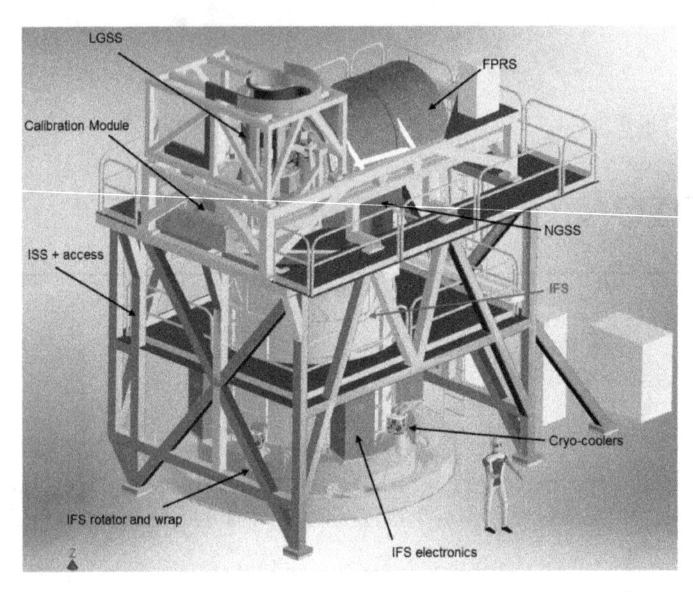

Fig. 9.1. The scale of the HARMONI instrument for the Extremely Large Telescope, compared to the size of a person. Reproduced with permission from HARMONI consortium.

9.1 Micro-Spectrographs

The simplest manifestation of this idea is to use a photonic lantern to feed a miniature version of a conventional spectrograph, such as sketched in Figure 9.2. The overall size of such a spectrograph can be very small, e.g. Figure 9.3, and off-the-shelf lenses and diffraction gratings often have satisfactory performance. These can often be housed in 3D printed mounts, since the tolerances are relaxed, leading to very cheap instruments. Robertson and Bland–Hawthorn (2012) gives a simple derivation of the design parameters for such an instrument, which we reproduced in Section 6.4.1. Micro-spectrographs can accept multiple fibre inputs arranged as a pseudo-slit; prototypes with 1×7 and 1×19 fibre photonic lanterns have been demonstrated Betters *et al.* (2013, 2014).

One particular advantage of a diffraction limited spectrograph fed with a single-mode fibre is the remarkably low amount of scattering generated. Recall from the discussions in Chapter 8 that scattering, especially in the presence of bright atmospheric emission lines, can severely compromise accurate sky subtraction. However, a diffraction limited spectrograph fed with a single-mode fibre displays remarkably little scattering, as shown in

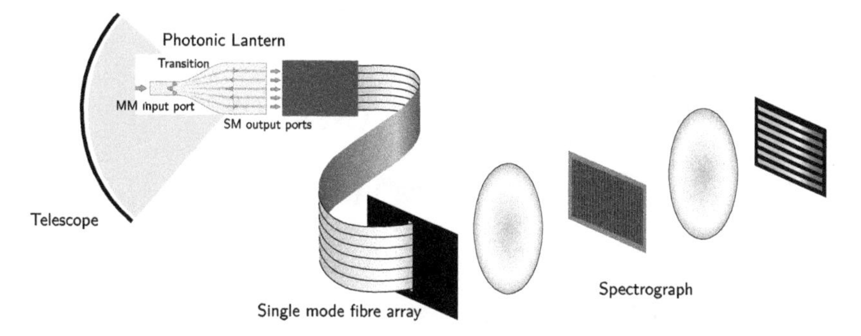

Fig. 9.2. Schematic diagram of a micro-spectrograph which uses a photonic lantern to divide the beam from the telescope into single-mode fibres, which can therefore feed a very compact diffraction limited spectrograph. Reproduced with permission from Betters *et al.* (2013). © The Optical Society.

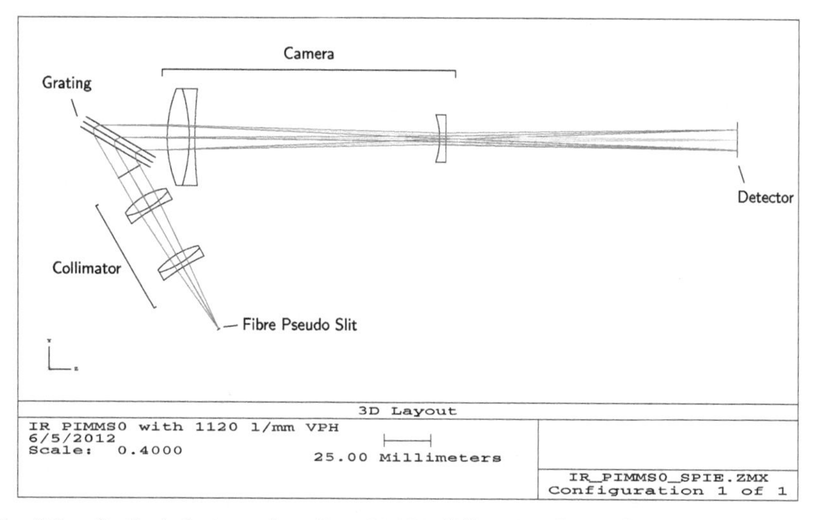

Fig. 9.3. Optical design of an $R = 31{,}000$ diffraction limited spectrograph. The diameter of each collimator lens is 25 mm, and the whole instrument fits within a 450 mm × 190 mm area. Reproduced with permission from Betters *et al.* (2012).

Figure 9.4, which compares the line spread function of the diffraction limited spectrograph of Betters *et al.* (2013), to conventional slit based spectrographs (IRIS2), and fibre-based spectrographs (AAOmega). These data are compared to theoretical line spread functions based on the illumination of the diffraction grating.

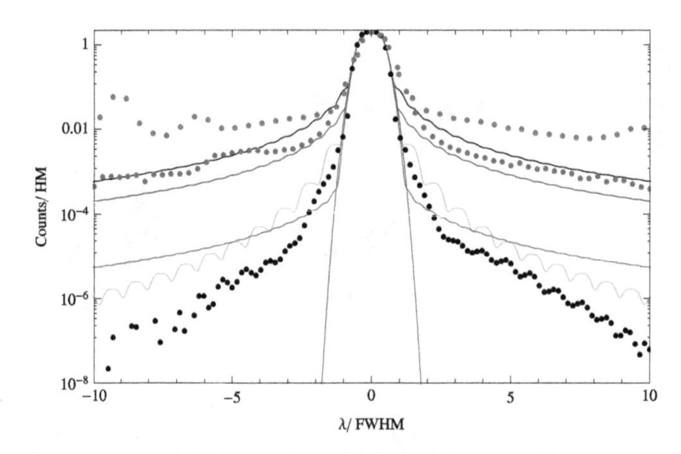

Fig. 9.4. Comparison of the logarithm of the PSFs due to different aperture apodisation functions, top-hat (black line), triangle (green line), Gaussian (red line), truncated Gaussian (blue line), wide-truncated Gaussian (purple line) and comparison to the measured PSF from IRIS2 (red points), AAOmega+2dF (blue points) and the diffraction limited spectrograph of Betters *et al.* (black points). The scaling of the data is such that the full width at half-maximum (FWHM) of each curve = 1, and the half-maximum on the y-axis = 1.

9.2 Arrayed Waveguide Gratings

One can miniaturise spectrographs even further by replacing the diffraction grating with photonic technologies such as **arrayed waveguide gratings**, or **AWGs**.

The principle of an AWG is sketched in Figure 9.5. In this case, light from a single-mode fibre enters into a free propagation zone, consisting of material of constant refractive index. The light from the fibre therefore spreads out as it propagates through this region. At the end of the free propagation zone, it enters an array of waveguides. These waveguides are curved, such that each one is a length ΔL longer than the preceding waveguide. Thus, the waveguides induce a phase difference of $\frac{2\pi}{\lambda} n\Delta L$ between adjacent waveguides, where n is the effective index of the waveguides and λ is the wavelength. This can be compared to the phase difference induced by a conventional grating as shown in Figure 9.5(b), and indeed the behaviour of AWGs is nearly identical to that of diffraction gratings in many respects. Therefore, on output light of different wavelength will add together in phase at different angles, and thereby form a spectrum at the end of the second free propagation zone.

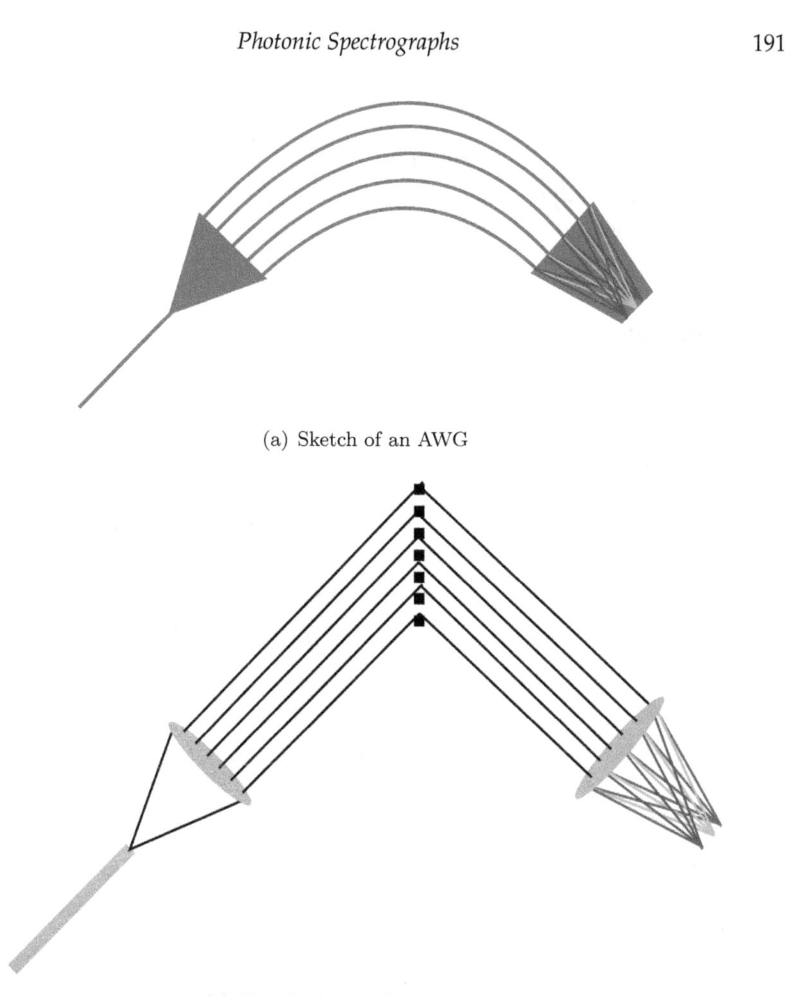

(a) Sketch of an AWG

(b) Sketch of an diffraction grating

Fig. 9.5. An arrayed waveguide grating: (a) works by imparting a phase difference into adjacent waveguides by means of a physically different path lengths. By comparison, a diffraction grating: (b) also imparts a phase difference by means of a physically different path length, due to reflections from adjacent facets.

One important way in which AWGs differ from conventional spectrographs is that there can be an arbitrary phase delay between adjacent waveguides set by their length difference. This effectively allows the AWG to operate in a high spectral order, without the necessity to operate at a large blaze angle. AWGs can be incredibly compact and still achieve very high resolving power. See, e.g. Allington–Smith and Bland–Hawthorn (2010) for a thorough discussion of this advantage.

9.2.1 *Theory*

Let us now examine this behaviour in more detail. Consider the light at output from two adjacent waveguides, as shown in Figure 9.6.

Recall the grating equation given in Section 6.4.1 (Equation 6.22). For an AWG we must modify this to account for the extra path length, q, between adjacent waveguides (this includes both the path length due to the difference in the lengths of the waveguides, and the path length difference due to the paths across the free propagation zone, see Section 9.2.1.1),

$$b \sin \theta + b \sin \theta_{\mathrm{i}} = m\lambda + q. \tag{9.2}$$

Differentiating this gives the same angular dispersion relation as for a conventional grating, given by Equation 6.23. However, now in the Littrow configuration, we have

$$2b \sin \theta_{\mathrm{i}} = m\lambda_{\mathrm{c}} + q, \tag{9.3}$$

and by the symmetry of the AWG $\theta_{\mathrm{i}} = \theta = 0$ so the angular and linear dispersion are then

$$\Delta\theta = -\frac{q\Delta\lambda}{b\lambda_{\mathrm{c}}}, \tag{9.4}$$

$$\Delta x = f_{\mathrm{AWG}} \frac{q\Delta\lambda}{b\lambda_{\mathrm{c}}}, \tag{9.5}$$

where f_{AWG} is the focal length of the AWG, and in the last line we have ignored the sign. Since AWGs must use single-mode waveguides, and

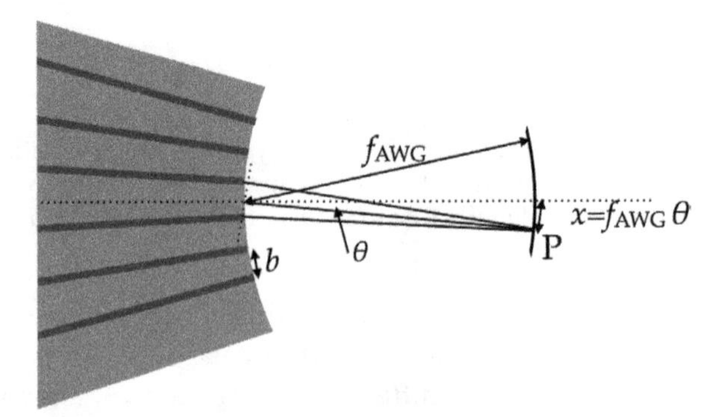

Fig. 9.6. Schematic of the output of an arrayed waveguide grating. A spectrum is formed on a circular surface at radius f_{AWG}, centred on the midpoint of the start of the free propagation zone. The output waveguides are also arranged on a circular arc of radius f_{AWG}, centred on the centre of the spectral surface.

recalling from Section 6.4.2 that the resolution of Gaussian beam may be approximated by 1.415 w_0 where w_0 is the beam waist in the image plane, the resolving power of the AWG can be expressed as

$$R_{\mathrm{AWG}} = \frac{\lambda_c}{\Delta\lambda} = q f_{\mathrm{AWG}} \frac{1}{1.415 w_0 b}. \tag{9.6}$$

Therefore, the resolving power can be arbitrarily high by selecting an appropriate value of q, as long as this does not introduce losses from overly small bend radii in the waveguides.

9.2.1.1 *Intensity distribution*

We can derive the intensity distribution from an AWG following the same procedure as for a diffraction grating. In this case, we will make the simplifying assumption that $f_{\mathrm{AWG}} \gg x$ and θ is small such that $\sin\theta \approx \theta$.

Let the difference in length between adjacent waveguides be ΔL. Therefore the difference in phase between exiting adjacent waveguides is

$$\Delta\phi_{\mathrm{wg}} = \frac{2\pi}{\lambda} n_{\mathrm{wg}} \Delta L, \tag{9.7}$$

where n_{wg} is the effective index of the waveguides, which is assumed to be equal for all waveguides. Added to this, there is another difference in phase due to the different path lengths travelled through the free propagation zones from the input fibre to the two waveguides under consideration, and from the output of the two waveguides to the point, P, located at an angle θ as in Figure 9.6.

The spectrum from the waveguides is formed on a circular surface of radius $r = f_{\mathrm{AWG}}$, centred at the midpoint of the output waveguides, which are themselves arranged on a circular surface of radius r, centred on the midpoint of the spectrum surface.

Therefore, the distance from the first waveguide to the point P is

$$z_1 = r + \frac{b}{2}\sin\theta \approx r + \frac{bx}{2r}. \tag{9.8}$$

Likewise, the distance from the second waveguide to the point P is

$$z_2 \approx r - \frac{bx}{2r}. \tag{9.9}$$

Therefore, the phase difference between adjacent waveguides due to the different paths across the free propagation zone is

$$\begin{aligned}\Delta\phi_{\text{fpz}} &= \frac{2\pi}{\lambda} n_{\text{fpz}} (z_1 - z_2), \\ &\approx \frac{2\pi}{\lambda} n_{\text{fpz}} \frac{bx}{r}.\end{aligned} \tag{9.10}$$

In general, an AWG can be fed by an input array of fibres, in which case there will be a similar phase difference on input. For simplicity we will only consider the case of injection by a single fibre, the output of which is equidistant from the input of the arrayed waveguides, and hence there is no extra phase difference on input.

In this case, the total phase difference for adjacent waveguides is

$$\begin{aligned}\Delta\phi &= \Delta\phi_{\text{wg}} + \Delta\phi_{\text{fpz}}, \\ &\approx \frac{2\pi}{\lambda} \left(n_{\text{wg}} \Delta L + n_{\text{fpz}} \frac{b\,x}{r} \right).\end{aligned} \tag{9.11}$$

There will be constructive interference at point P when $\Delta\phi = 2\pi m$ and m is an integer, hence

$$\lambda(x) \approx \frac{1}{m} \left(n_{\text{wg}} \Delta L + n_{\text{fpz}} \frac{b\,x}{r} \right). \tag{9.12}$$

If we make the simplifying assumptions that $n_{\text{wg}} = n_{\text{fpz}}$, and further that the phase difference between adjacent waveguides is an integer multiple of 2π, i.e. $\Delta L = \frac{\lambda_0}{n_{\text{wg}}}$, where λ_0 is the central wavelength, then we have

$$\lambda(x) \approx \frac{\lambda_0}{m} + \frac{b\, n_{\text{fpz}}}{m\, r} x. \tag{9.13}$$

If the electric field amplitude arriving at the point P from the first waveguide is E_0, then the total electric field at point P from N waveguides is given by

$$\begin{aligned}E_{\text{P}} &= E_0 \sum_{j=0}^{j=N-1} \mathrm{e}^{\mathrm{i}j\Delta\phi}, \\ &= E_0 \left(\frac{\mathrm{e}^{\mathrm{i}N\Delta\phi} - 1}{\mathrm{e}^{\mathrm{i}\Delta\phi} - 1} \right).\end{aligned} \tag{9.14}$$

Therefore, the intensity at point P is given by

$$\begin{aligned} I_{\mathrm{P}} &= |E_{\mathrm{P}}|^2, \\ &= I_0 \frac{\sin^2 \frac{N\Delta\phi}{2}}{\sin^2 \frac{\Delta\phi}{2}}, \end{aligned} \tag{9.15}$$

where $I_0 = E_0^2$.

Example 9.1: Derive Equation 9.15.
We have

$$\begin{aligned} I_{\mathrm{P}} &= |E_{\mathrm{P}}|^2, \\ &= E_0^2 \left(\frac{\mathrm{e}^{\mathrm{i}N\Delta\phi} - 1}{\mathrm{e}^{\mathrm{i}\Delta\phi} - 1}\right)\left(\frac{\mathrm{e}^{-\mathrm{i}N\Delta\phi} - 1}{\mathrm{e}^{-\mathrm{i}\Delta\phi} - 1}\right), \\ &= E_0^2 \left(\frac{2 - \mathrm{e}^{\mathrm{i}N\Delta\phi} - \mathrm{e}^{-\mathrm{i}N\Delta\phi}}{2 - \mathrm{e}^{\mathrm{i}\Delta\phi} - \mathrm{e}^{-\mathrm{i}\Delta\phi}}\right). \end{aligned}$$

And recalling $\cos\theta = (\mathrm{e}^{\mathrm{i}\theta} + \mathrm{e}^{-\mathrm{i}\theta})/2$,

$$I_{\mathrm{P}} = E_0^2 \left(\frac{2 - 2\cos N\Delta\phi}{2 - 2\cos \Delta\phi}\right),$$

which, upon substituting $\cos 2\theta = \cos^2\theta - \sin^2\theta$, becomes

$$\begin{aligned} I_{\mathrm{P}} &= E_0^2 \left(\frac{1 - \cos^2 \frac{N\Delta\phi}{2} + \sin^2 \frac{N\Delta\phi}{2}}{1 - \cos^2 \frac{\Delta\phi}{2} + \sin^2 \frac{\Delta\phi}{2}}\right), \\ &= I_0 \frac{\sin^2 \frac{N\Delta\phi}{2}}{\sin^2 \frac{\Delta\phi}{2}}, \end{aligned}$$

where in the last line we have used $\sin^2\theta + \cos^2\theta = 1$. Q.E.D.

Now, taking the case in which the outputs from each waveguide are in phase prior to crossing the free propagation zone, i.e. $\Delta\phi_{\mathrm{wg}} = 0$ and $\Delta\phi = \Delta\phi_{\mathrm{fpz}}$

$$I_{\mathrm{P}} = I_0 \frac{\sin^2 \frac{\pi N n_{\mathrm{fpz}} b x}{\lambda r}}{\sin^2 \frac{\pi n_{\mathrm{fpz}} b x}{\lambda r}}, \tag{9.16}$$

which is sketched in Figure 9.7 for $N = 2, 3, 5$ and 10.

9.2.1.2 *Free spectral range*

The intensity distribution in Equation 9.16 sketched in Figure 9.7 is made up of two parts: a rapidly varying sinusoid described by the numerator and a slowly varying envelope described by the denominator. The maxima occur when the denominator is zero, i.e. when

$$\begin{aligned}\frac{\pi n_{\mathrm{fpz}} b x_{\max}}{\lambda r} &= m\pi, \\ \implies x_{\max} &= m\frac{\lambda r}{n_{\mathrm{fpz}} b}, \end{aligned} \tag{9.17}$$

and m is an integer.

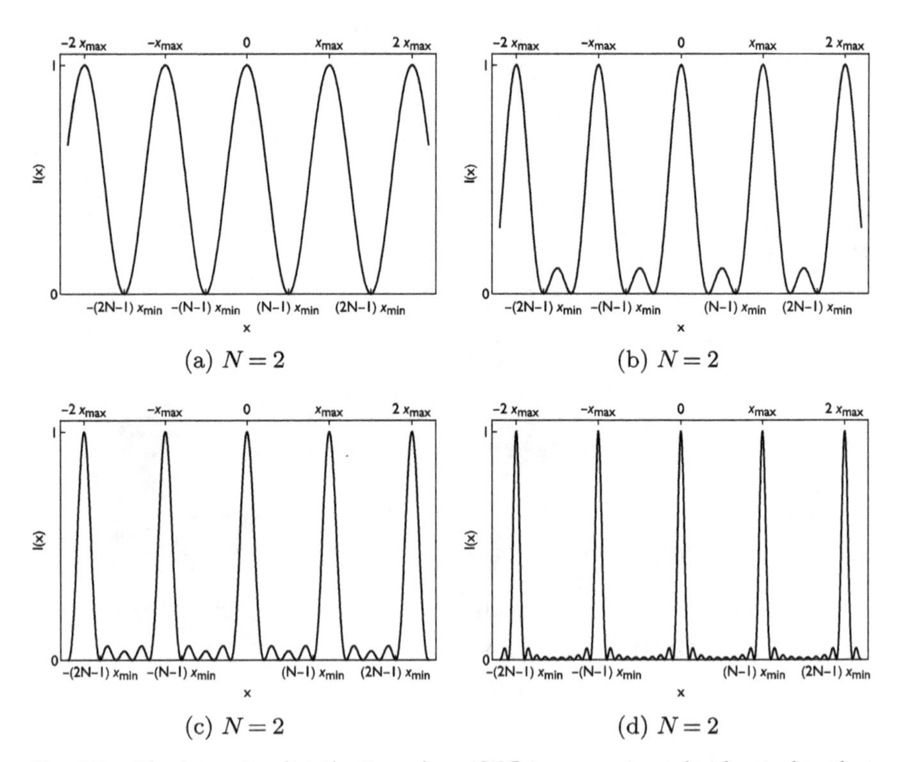

(a) $N = 2$ (b) $N = 2$

(c) $N = 2$ (d) $N = 2$

Fig. 9.7. The intensity distribution of an AWG is approximately identical to that of a diffraction grating. The resolving power increases with the number of waveguides N, and the ticks mark the locations of the principal maxima and minima.

The minima occur when the numerator is zero, except when that coincides with a maximum, i.e.

$$\frac{\pi N n_{\text{fpz}} b x_{\text{min}}}{\lambda r} = p\pi,$$
$$\implies x_{\text{min}} = p \frac{\lambda r}{N n_{\text{fpz}} b}, \tag{9.18}$$

where $p = 1, 2, \ldots, N-1, N+1, \ldots,$ etc.

Differentiating Equation 9.17

$$\frac{\mathrm{d}\lambda}{\mathrm{d}m} = -\frac{x_{\text{max}} n_{\text{fpz}} d}{m^2 r}, \tag{9.19}$$

which, when $\mathrm{d}m = 1$, gives

$$\Delta\lambda = \frac{\lambda}{m}. \tag{9.20}$$

9.2.2 *Development of AWGs for astronomy*

The potential of AWGs for astronomical spectroscopy was first pointed out by Bland–Hawthorn and Horton (2006). The first on-sky tests were carried out by Cvetojevic *et al.* (2009) at the Anglo-Australian Telescope. In order to make the AWG tested compatible with astronomical requirements several modifications were made. AWGs for mode division multiplexing in telecomms. have an output array of single-mode fibres, positioned to accept light of particular wavelengths from the image plane. For astronomical purposes, a continuous spectrum, rather than a discrete set of signals, is required, and so the output fibres were removed, and the free-propagation zone was polished flat to produce an optical quality surface. Note that this will introduce some defocussing along the surface, since the focal plane really follows a circular surface, as in Figure 9.6. Second, for the first tests, the AWG was packaged with just one single-mode fibre input. This device is shown in Figure 9.8.

Following this, AWGs were further developed for astronomy, increasing the resolving power and number of fibre inputs (Cvetojevic *et al.* 2012b). Note, an individual AWG can accept multiple adjacent SMF inputs, resulting in overlapping spectra at the focal plane. These can then be separated by cross-dispersing the output, before recording the spectra. These developments led to the first stellar spectra recorded with an AWG, again at the AAT, with the AWG retrofitted to feed an existing spectrograph delivering a resolving power of $R = 2{,}500$ at a central wavelength of 1.6 μm (Cvetojevic *et al.* 2012a).

Fig. 9.8. Photograph of the first AWG prototype to be tested on-sky for astronomical purposes, with a single fibre input. Reproduced with permission from Cvetojevic *et al.* (2009). © The Optical Society.

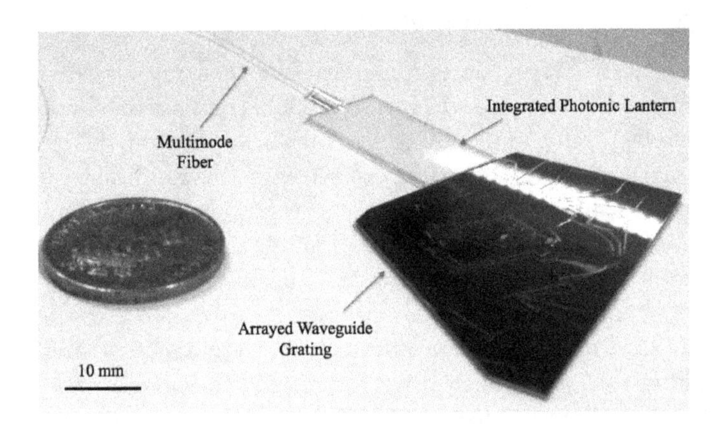

Fig. 9.9. Example of a packaged AWG from. This device includes a multimode input, and photonic lantern with 37 single-mode outputs feeding an AWG. Reproduced with permission from Cvetojevic *et al.* (2017). © The Optical Society.

These early results demonstrate the potential of AWGs for astronomy, but several further improvements are necessary to make an astronomically useful instrument. In particular, the AWG efficiency, the coupling efficiency between the input fibres and the waveguides, polarisation sensitivity, and modal noise effects from the photonic lantern injection are all being actively studied (Cvetojevic *et al.* 2017; Gatkine *et al.* 2016, 2017). Figure 9.9 shows an example of a packaged AWG from Cvetojevic *et al.* (2017).

9.3 Photonic Echelle Gratings

A different type of micro-spectrograph is the photonic echelle grating, which was first proposed for astronomical use by Watson (1995). These devices are conceptually very similar to a standard echelle reflection grating, but the grating is written into a slab waveguide as in Figure 9.10.

The advantage of using a photonic echelle grating over an AWG is that the number of reflecting facets can be very much larger than the number of individual waveguides in an AWG. Therefore, the resolving power $R = mN$ can be correspondingly larger. Furthermore, they can be used directly with few-moded fibres without loss of performance relevant to astronomical applications. A good review of the requirements on photonic echelles for astronomy is given by Bland-Hawthorn and Horton (2006).

Ironically, despite being one of the very first integrated photonic devices suggested for astronomy (Watson 1996), and being seemingly more suited and more easily adapted to astronomy, they have attracted less attention because they have proved harder to adapt to the needs of telecommunications (Pathak *et al.* 2014), especially as regards polarisation dependence and birefringence and very strict requirements on cross-talk (Bland-Hawthorn and Horton 2006), and therefore, manufacturing

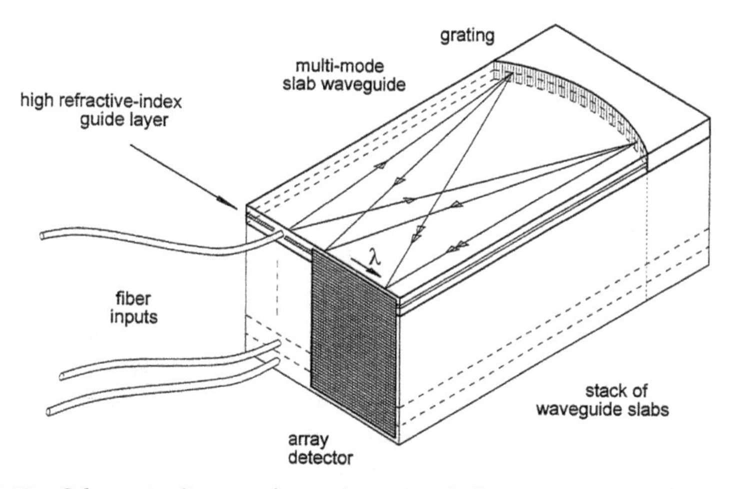

Fig. 9.10. Schematic diagram for a photonic echelle spectrograph, showing the principle of operation. Each layer of the stack contains a slab waveguide, into which is written an echelle grating. The light reflected from will form a spectrum at the image plane, which could then either be directly imaged as in the diagram, or reimaged onto a detector. Reproduced with permission from Watson (1995).

methods lag behind those of AWGs. A notable exception is the recent work by Stoll *et al.* (2020) who have investigated various designs for PEGs for astronomical spectroscopy.

9.4 Lippmann Spectrograph

A rather different scheme for a micro-spectrograph is to use a Lippmann spectrograph, or stationary wave integrated Fourier transform spectrograph (SWIFTS), embedded into a waveguide. In this technique, light is either injected into a single-mode waveguide and reflected at one end, or is injected into both ends of a single-mode waveguide. In either case, this results in two identical and counter-propagating waves, which will construct a standing wave inside the waveguide. This standing wave can be detected by placing nano-dots along the outside of the waveguide which scatter the evanescent field onto a detector, recording an interferogram. The interferogram can then be analysed in a similar way to a Fourier transform spectrometer in order to recover the spectrum. These processes are illustrated in Figure 9.11.

This birth of this technique can be traced to a method of colour photography developed by Gabriel Lippmann in the 1890s, for which he won the 1908 Nobel Prize (Lippmann 1891, 1894). This idea was resurrected by Connes and Le Coarer (1995) in light of new developments in photonics and detector technology, and was developed into a working device by Etienne Le Coarer and his colleagues (see, e.g. Le Coarer *et al.* 2007 for a thorough overview, or Blind *et al.* 2017 for a more recent review).

The primary difference between Lippmann spectroscopy and the other techniques described above is that in a conventional diffraction grating spectrograph or an AWG the light is dispersed such that each pixel sees a different wavelength. On the other hand, in a Lippmann spectrograph all pixels see a contribution from all wavelengths. Unfortunately for a continuum source this is disadvantageous for signal-to-noise. Recall Equation 3.1 describing the signal-to-noise of an observation. Consider an AWG and a Lippmann spectrograph, each observing a continuous source, both of which have the same resolving power, and so record the same number, M, of resolution elements, $\Delta\lambda$. Now, if the observations are background limited the signal-to-noise per resolution element for an AWG is approximately

$$\mathrm{SNR}_{\mathrm{AWG}_{\mathrm{sky}}} = \frac{N_{\mathrm{S}}}{\sqrt{B}}, \tag{9.21}$$

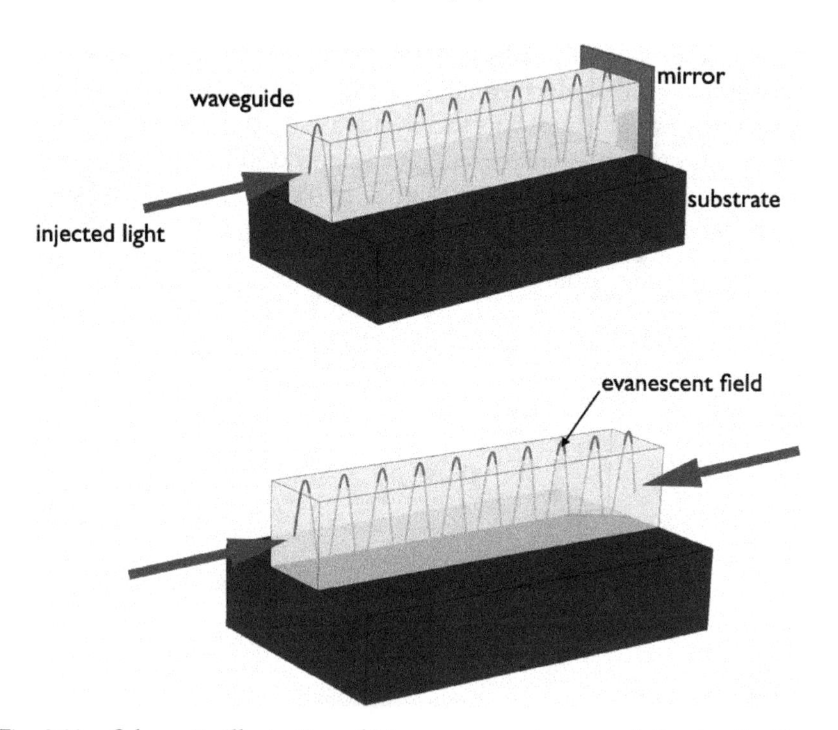

Fig. 9.11. Schematic illustration of Lippmann spectroscopy following Le Coarer *et al.* (2007). The top figure shows the Lippmann technique with a mirror at the end of the waveguide; the bottom figure shows the counter-propagation techniques in which the input field is divided and injected into either end of the waveguide.

where here we take N_S to be the counts in the signal per resolution element, and B to be the background counts per resolution element. For a Lippmann spectrograph the signal and background are spread over the M resolution elements, and the signal-to-noise must be recovered by summing together the contribution from each resolution element

$$\begin{aligned}
\mathrm{SNR}_{\mathrm{Lipp_{sky}}} &= \sqrt{\sum_{m=1}^{M}\left(\frac{\frac{N_S}{M}}{\sqrt{\frac{B}{M}}}\right)^2}, \\
&= \frac{N_S}{\sqrt{MB}}, \\
&= \frac{1}{\sqrt{M}}\mathrm{SNR}_{\mathrm{AWG_{sky}}}.
\end{aligned} \tag{9.22}$$

So the signal-to-noise is less by a factor $\sqrt{M}$ in a Lippmann spectrograph. However, for sources which contain only a few emission lines, and therefore, require fewer spectral bins this disadvantage is ameliorated.

Since a Lippmann spectrograph is comprised of one or more waveguides it is readily integrable with other photonic components, such as photonic lanterns.

CHAPTER 10

WAVELENGTH CALIBRATION

Accurately calibrating the wavelength response of a spectrograph is an important part of astronomical spectroscopy. This is typically done by intermittently, or sometimes simultaneously, observing a source with emission lines of known wavelength, such as a gas-discharge (arc) lamp, or more recently laser combs. The accuracy required for wavelength calibration depends upon the science goals, with some typical values listed in Table 10.1. The values are given in metres per second, i.e. $\Delta v = c\frac{\Delta\lambda}{\lambda}$. Note that it is not necessary to observe with a resolving power of $\Delta\lambda/\lambda$ in the previous equation; the centroid of a particular spectral feature can be measured to an accuracy of $\sim\frac{\Delta\lambda}{\mathrm{SNR}}$, where SNR is the signal-to-noise, according to the Cramér–Rao bound. Thus, with a high signal-to-noise, the wavelength of a particular feature can be measured to much better accuracy than the resolution; also, many features can be measured simultaneously, further increasing the accuracy.

Astrophotonics is important for wavelength calibration in two ways. First, due to their small sizes, photonic devices can be more easily stabilised, e.g. against temperature shifts, and therefore can be made more robust against systematic errors. Second, photonic devices can be used to provide accurate wavelength sources. Both these will be discussed as follows.

Table 10.1. Typical accuracies in the required velocity measurements for example science cases.

Science case	Accuracy ($\mathrm{m\,s^{-1}}$)
Galaxies, redshifts	600,000
Galaxy kinematics	100,000
Stellar astrophysics	15,000
Exoplanets — Hot Jupiters	50
Exoplanets — Earth-like planets	0.2

Example 10.1: What signal-to-noise is required to centroid a single feature at a wavelength of 800 nm to an accuracy of 10 m s^{-1} if the resolving power is $R = 50{,}000$?
For $R = 50{,}000$ and $\lambda = 800$ nm, $\Delta\lambda = 800/50{,}000 = 0.016$ nm. However, the required accuracy is $800 \times \frac{10}{3\times10^8} = 2.7\times10^{-5}$ nm. Therefore, the required signal-to-noise is $\mathrm{SNR} \sim \frac{0.016}{2.7\times10^{-5}} = 600$.

10.1 Stability

In theory, it is possible to improve the accuracy of the wavelength calibration (or indeed any measurement) with greater signal-to-noise, e.g. obtained by a large number of repeated measurements. In practice, there will be a limit to the accuracy due to systematic errors. In the case of wavelength calibration for a spectrograph, these will usually be due to the instability of the instrument. For example, temperature variations can cause thermal expansion of the mechanical housings, pressure variations can cause turbulence and deviations in the light path, and gravity variations (for an instrument attached to the telescope) can cause flexure.

These considerations hold equally true for many astrophotonic instruments. Indeed, many of the devices discussed in this book have other uses as sensors. For example, the resonant wavelengths of a ring resonator will change with temperature, due to thermal expansion and changes in refractive index; likewise, the Bragg wavelengths of a fibre Bragg grating may also change due to an induced strain.

Nevertheless, the small size of many of these devices makes it relatively easy to control or compensate any such changes. For example, it is much easier to control the temperature of an arrayed waveguide grating with a size of a few centimetres, than a spectrograph of many metres in size.

In some cases, the stabilisation can be built into the device itself. Fibre Bragg gratings are very sensitive to changes in temperature, and indeed can be used as temperature sensors, as the reflected wavelengths will change with temperature. However, FBGs are also sensitive to strain on the fibre, which similarly makes them useful as strain gauges. Together, these two effects can be used to make athermal devices, as was done for the FBGs used for OH suppression described in Chapter 8. The FBG is packaged in a metal tube, the thermal expansion of which adds a strain to the fibre which exactly compensates the change in Bragg wavelength due to the temperature change. Figure 10.1 shows the change in reflected wavelength as a function of temperature for both an athermally packed FBG and an uncompensated FBG, from Yoffe *et al.* (1995).

In other cases, the stability to temperature variations must be actively maintained. The resonant wavelength of a ring resonator is sensitive to temperature changes due to both thermal expansion and changes in refractive index. These can be controlled with micro-heaters or stress-optic modulators embedded into the photonic chip itself, underneath the rings;

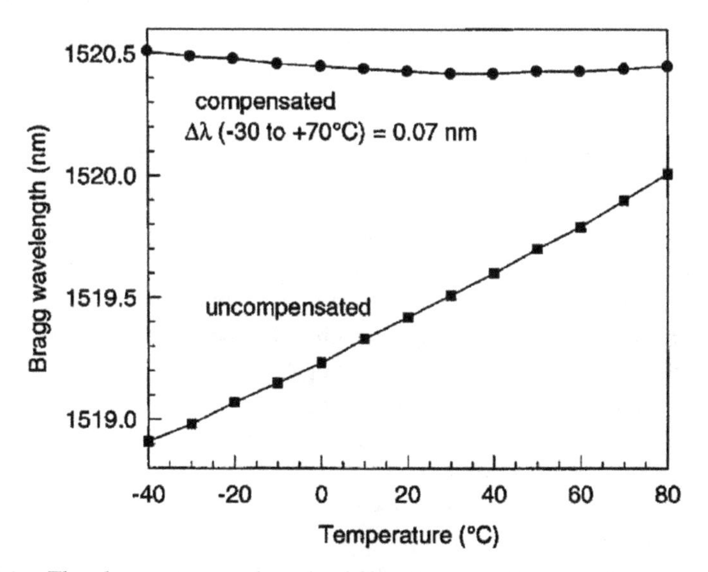

Fig. 10.1. The change in wavelength of fibre Bragg grating as a function of temperature, for uncompensated and strain-compensated devices. Reproduced with permission from Yoffe *et al.* (1995). © The Optical Society.

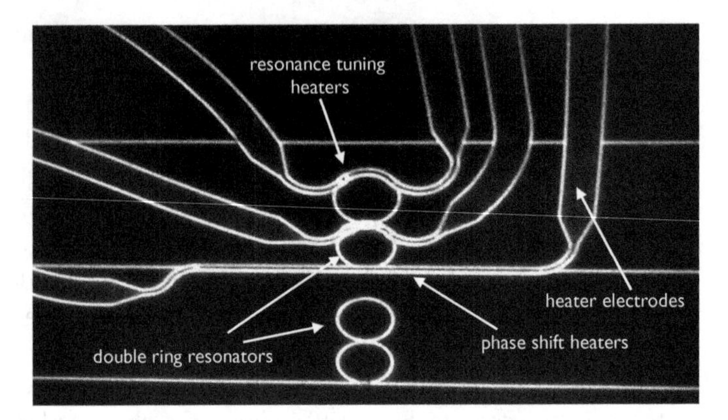

Fig. 10.2. Vernier-coupled double-ring resonator with heaters for tuning the resonance and coupling, manufactured by LioniX International.

see Figure 10.2. Changes can be monitored, e.g. with a wavemeter measuring the light from a drop port, and sending feedback to the heater in real time, whilst making a long exposure of the light exiting the through port. Likewise, the phase of the light in the coupling region can also be modulated using a heater to fine-tune the coupling coefficients. Since the resonant wavelength of ring resonators usually needs to be tuned (known as trimming) post manufacture in any case, adding an active tuning heater or something similar is generally a good idea.

10.2 Wavelength Sources

The second way in which astrophotonics plays an important role in wavelength calibration is in generating the calibration source itself. Traditionally, wavelength calibration has been achieved using arc discharge lamps. These provide a source of emission lines whose absolute wavelength is known to very high accuracy. However, the intensities of different lines from the same lamp can be very different, as can the line density, see, e.g. Figure 10.3. This can make it problematic to achieve very accurate wavelength calibration at any wavelength and high resolving power, as in certain cases, the brightest lines may be saturated, or else the faintest lines too faint.

This problem is now being addressed with the use of frequency combs for high-resolution spectrographs. These provide a comb of emission lines of comparable intensity, equally spaced in frequency. The most precise of

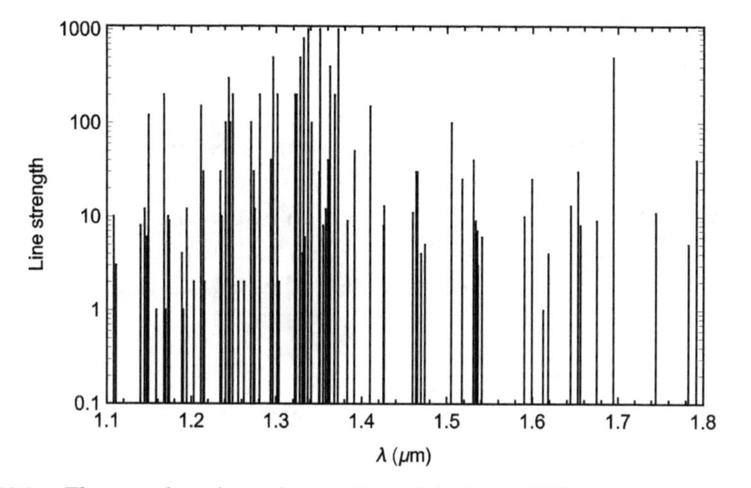

Fig. 10.3. The wavelengths and intensities of Ar in the NIR region, as commonly used for wavelength calibration with an arc lamp.

these devices use a mode-locked laser, which is locked in frequency to an atomic clock. We do not discuss these frequency combs in this book, focussing instead on astrophotonic combs using waveguides; the interested reader is directed to Murphy *et al.* (2007) and Steinmetz *et al.* (2008), and works citing these.

Ring resonators (Section 8.4) provide several means of generating frequency combs, using different phenomena, which provide different levels of precision, albeit at a cost of increasing complexity. At the most basic level, a ring resonator can provide a comb of frequencies by simply including a drop port, as in Figure 8.16. The resonant frequencies of the ring will evanescently couple into the drop port, and provide a series of lines at wavelengths $\lambda = n_{\mathrm{eff}}L/m$, from Equation 8.5. The output of a simple add/drop resonator can be found following a similar analysis to that presented in Section 8.4. The free spectral range is determined purely by the ring, is proportional to the reciprocal of the circumference, and is identical to that of a notch filter given by Equation 8.13. Ring resonators used in this way can provide a compact and simple wavelength calibration source (Ellis *et al.* 2012b, 2017; Lee *et al.* 2012). By tuning the circumference of the ring appropriately, one may select a wide range of FSRs, e.g. from 0.02 to 20 nm, thus enabling the appropriate line density for the spectrograph. In comparison, laser frequency combs must usually be filtered subsequently, e.g. using a Fabry–Pérot etalon, since the inherent line density is too high for useful purposes.

Example 10.2: Consider a ring resonator in Si_3N_4 with a group index of $n_g = 1.9$. What radius must the ring have to calibrate an $R = 10{,}000$ spectrograph with a line density of $10 \times$ FWHM working at $\lambda = 1.5$?
The FWHM is $1.5/10{,}000 = 1.5 \times 10^{-4}$ μm, and therefore we require an FSR $= 1.5 \times 10^{-3}$ μm. From Equation 8.13, the FSR is given by

$$\Delta\lambda = \frac{\lambda^2}{Ln_g}, \tag{10.1}$$

which, solving from $L = 2\pi R$, gives $R = 125.6$ μm.

Although adequate for moderate precision, constructing a frequency comb from a ring resonator in this way is not sufficient for very high accuracy, as the comb spacing is not fixed, as the group index, n_g, is a function of wavelength, see Equation 4.33. Furthermore, there may be significant light between the resonant lines, depending on the Q factor of the ring.

A more precise frequency comb can be generated from a ring resonator by exploiting nonlinear effects in the waveguide material, in which the polarisation density of the waveguides is changed by the electric field within it. Consider a ring resonator with a resonant frequency f_0. If this ring is fed with an intense laser at a slightly different frequency, f_1, then a third frequency at a wavelength $f_2 = 2f_1 - f_0$ will be generated, due to a process known as **degenerate four-wave mixing**. This process can then be repeated due to the mixing of the three frequencies, and so on in a cascade process to generate a frequency comb, see Figure 10.4.

Potential advantages of ring resonator frequency combs for astronomy are that they can be generated with any FSR (as described above), and at any central wavelength over a broad wavelength range, and in a compact energy-efficient manner. Furthermore, due to the CMOS-compatible lithographic manufacturing process, they are easily replicated and produced. Ring resonator based frequency combs are now being applied in high-resolution spectrographs (e.g. Obrzud *et al.* 2019; Suh *et al.* 2019).

Another very compact type of photonic comb is the fibre Fabry–Pérot etalon (Halverson *et al.* 2014; Schwab *et al.* 2015). This works on the same principle as a Fabry–Pérot etalon. Two single-mode fibres are spliced to a pair of parallel mirrors, as in Figure 10.5. The mirrors are of high reflectivity, so the light will undergo multiple reflections between them. Thus,

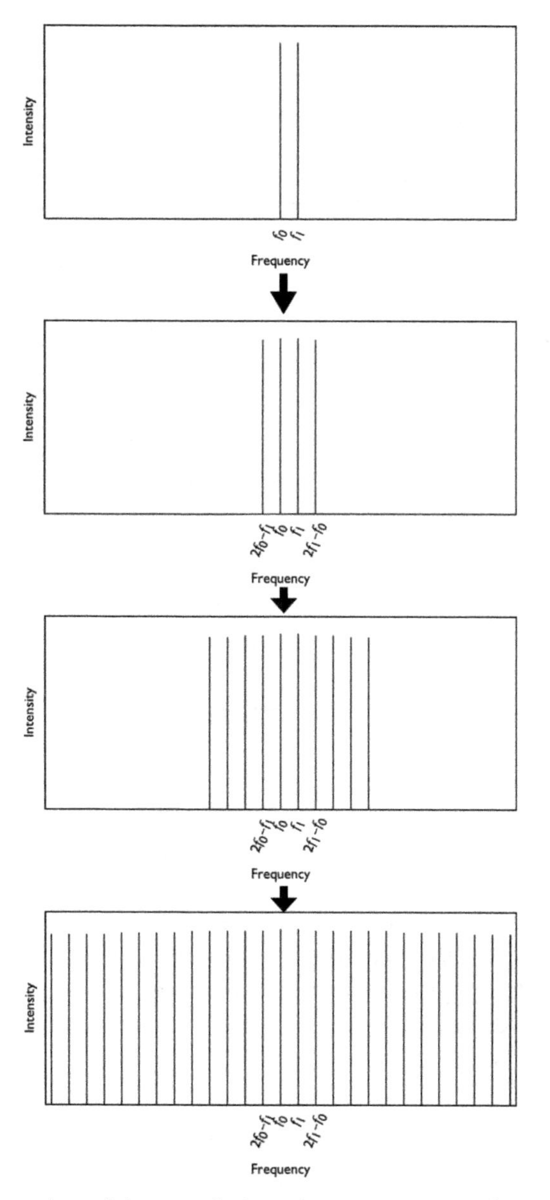

Fig. 10.4. Illustration of the population of a ring resonator frequency comb by four-wave mixing. A ring, with resonant frequency f_0, is pumped with a laser at frequency f_1. This then generates two new frequencies at $2f_1 - f_0$ and $2f_0 - f_1$. These four frequencies then combine in the same pairwise manner, and so on, until a frequency comb is generated.

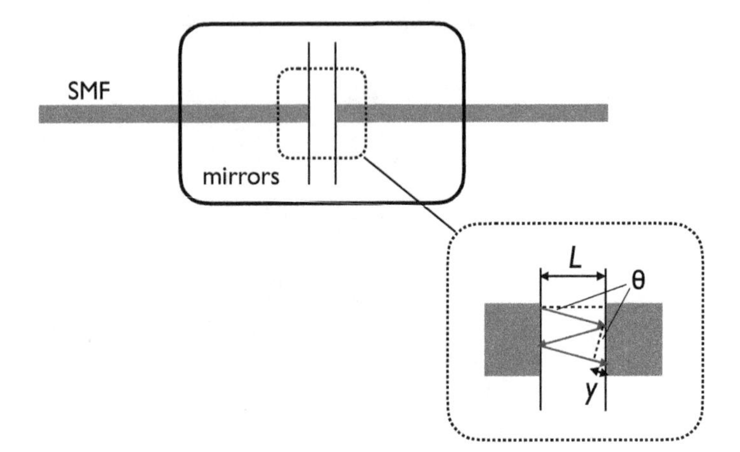

Fig. 10.5. A sketch of a fibre Fabry–Pérot etalon.

there will be a significant build-up of light for those wavelengths that add together in phase after multiple reflections, i.e.

$$m\lambda \approx 2nL, \tag{10.2}$$

where m is an integer, L is the length of the cavity, and n is the refractive index of the material in the cavity (cf. Section 8.4). Thus, a frequency comb consisting of these particular wavelengths will be transmitted to the exit fibre.

Equation 10.2 is only approximate, as the light may not travel directly along the optical axis, but could reflect at some angle θ to this, as shown in the inset sketch in Figure 10.5. From the figure, we can see that the phase difference between successive rays exiting the cavity at an angle θ after two reflections is

$$\begin{aligned}\delta &= \frac{2\pi}{\lambda} n y + 2\phi \\ &\approx \frac{4\pi}{\lambda} n L \cos\theta,\end{aligned} \tag{10.3}$$

where ϕ is the phase change due to the reflection at the mirror, which is usually small and therefore negligible compared to d/λ, which is usually large. Now, if both surfaces have a reflectance of R, then the light exiting the cavity can be found from the sum of the complex amplitudes, i.e.

$$\begin{aligned} E_{\text{out}} &= E_{\text{in}} T \sum_{m=0}^{\infty} R^m \mathrm{e}^{\mathrm{i}(\phi_0 + m\delta)} \\ &= E_{\text{in}} T^2 \left(\frac{\mathrm{e}^{\mathrm{i}\phi_0}}{1 - R\mathrm{e}^{\mathrm{i}\delta}} \right), \end{aligned} \tag{10.4}$$

where $T = 1 - R$ is the transmittance (assuming no losses due to absorption) and ϕ_0 is the arbitrary output phase of the first ray. Therefore, the fractional intensity on output is

$$\frac{I_{\text{in}}}{I_{\text{out}}} = \left| \frac{E_{\text{out}}}{E_{\text{in}}} \right|^2 = \frac{T^2}{1 - 2R\cos\delta + R^2}. \tag{10.5}$$

Manipulating this expression further, and using the identity $\cos 2x = \cos^2 x - \sin^2 x = 1 - 2\sin^2 x$, we find

$$\frac{I_{\text{in}}}{I_{\text{out}}} = \frac{1}{1 + F\sin^2 \frac{\delta}{2}}, \tag{10.6}$$

where

$$F = \frac{4R}{(1-R)^2}. \tag{10.7}$$

Now, the FWHM of the comb can be found from the value of δ when Equation 10.6 =1/2, i.e.

$$\delta_{\text{HM}} = \pm 2\sin^{-1}\sqrt{\frac{1}{F}} \approx \frac{4}{\sqrt{F}}, \tag{10.8}$$

where the approximation is valid when R is high, in which case, F is large, and $\sin^{-1}\sqrt{1/F} = \sqrt{1/F}$. Or, in terms of wavelength,

$$\delta\lambda = \frac{\delta_{\text{FWHM}}}{\frac{d\delta}{d\lambda}} = \frac{\lambda^2}{\sqrt{F}\pi n_{\text{g}} L \cos\theta}, \tag{10.9}$$

where n_{g} is the group index (Equation 4.33), to take into account the wavelength dependence of n.

The free spectral range of the comb is the wavelength difference between adjacent orders. This occurs when the phase difference given by

Equation 10.3 is equal to $m2\pi$, where m is an integer, i.e.

$$\frac{4\pi}{\lambda} nL\cos\theta = m2\pi,$$

$$\implies \frac{dm}{d\lambda} = -\frac{2nL\cos\theta}{\lambda^2}, \tag{10.10}$$

and when the difference between orders $\Delta m = 1$, we have

$$\Delta\lambda = \frac{\lambda^2}{2n_g L\cos\theta}. \tag{10.11}$$

The finesse, $\mathcal{F}$, of the comb is the ratio of the free spectral range to the FWHM, i.e.

$$\mathcal{F} = \frac{\Delta\lambda}{\delta\lambda},$$

$$= \frac{\pi\sqrt{F}}{2}. \tag{10.12}$$

An interesting application of a fibre Fabry–Pérot comb has been to map the aberrations of a high-resolution fibre-fed multi-object spectrograph (Bland–Hawthorn *et al.* 2017). This is analogous to measuring the aberrations of an imager using a physical mask with small holes, illuminated by a lamp to produce a series of diffraction-limited spots on the detector, from which all aberrations and distortions can be measured. Comparable measurements in a spectrograph are typically extremely hard due to the dispersion in wavelength of the spots, but this has been elegantly solved by using a photonic comb produced by a fibre Fabry–Pérot. Since the FFP uses single-mode fibres, the output will be an approximately Gaussian PSF at wavelength of the comb. Measuring the actual PSF over the entire detector allows the measurement and characterisation of all the optical aberrations and distortions of the spectrograph.

PART III

FUTURE ASTROPHOTONICS

CHAPTER 11

UNEXPLOITED PHOTONICS

In Part II, we illustrated the breadth of potential of astrophotonics with examples of current instruments and research. The field of astrophotonics is currently at a watershed, with many technologies now emerging from the laboratory to be tested in prototype instruments and on-sky demonstrations, and the first facility class instruments using astrophotonics are now in use. In Table 11.1, we summarise the current situation, by giving an estimate of the Technology Readiness Level (TRL), as defined by the European Southern Observatory (adapted from the NASA TRLs) for the different astrophotonic technologies we have discussed.

Despite the growing number of successful instruments and prototypes, it should be emphasised that astrophotonics is still a nascent and developing field of research. Astrophotonics has only begun to exploit a few of the more obvious and easily adaptable photonic technologies; outside astronomy the field is huge and extremely diverse, even when used in our restricted definition of devices embedded into waveguides. We now turn our attention to the future possibilities for astrophotonics by examining those technologies which have not yet been considered or adapted for astronomy. Since the field of photonics is so large, we must content ourselves with a somewhat biassed selection of those technologies which we believe are most likely to play a future role in astronomy.

In the first place there are alternative astrophotonic solutions to the problems and applications already discussed. A good example is the concurrent investigation of ring resonators and fibre Bragg gratings for OH suppression (Chapter 8).

Table 11.1. Estimates of the approximate Technology Readiness Level of current astrophotonic technologies.

TRL	Meaning	Examples
1	Basic principles observed and reported	Laguerre–Gaussian mode sorting coronagraphy Photonic AO
2	Technology concept and/or application formulated	Photonic echelle gratings for spectroscopy
3	Analytical and experimental critical function and/or characteristic proof of concept	Ring resonator OH suppression
4	Component and/or breadboard validation in laboratory environment	
5	Component and/or breadboard validation in relevant environment	Single-mode spectrographs (e.g. PIMMS)
6	System/Subsystem model or prototype demonstration in a relevant environment	Arrayed waveguide gratings
7	System prototype demonstration in an operational environment	FBG OH suppression (e.g. GNOSIS and PRAXIS)
8	Actual system completed and qualified through test and demonstration	Photonic astrocombs, including ring resonators and fibre Fabry–Pérot etalons Aperture masking photonic reformatters (e.g. Dragonfly, VAMPIRES)
9	Actual system proven through successful mission operations	Interferometric beam combiners (e.g. PIONIER, FLUOR, GRAVITY)

Another example already discussed is the use of photonic echelle gratings for miniature spectrographs (Section 9.3). Although discussed in the literature (Bland–Hawthorn and Horton 2006; Watson 1995, 1996) there has been very little serious investigation of these promising devices in the laboratory (see Stoll *et al.* 2020 for recent progress).

However, we wish to focus our attention on already existing photonic devices, or those currently under development, which have not yet been discussed, and which have the potential to offer new solutions for astronomical instruments.

11.1 Unexploited Photonic Devices

Photonics is an enormous field of research; see for example proceedings of any large photonics conference, e.g. SPIE Photonics West, CLEO etc., or the breadth of topics covered in photonics journals such as Nature Photonics. Mining this field for all potentially useful astrophotonic devices is an overwhelmingly large task. Since we wish here to give a general overview rather than address solutions to a specific problem, we will focus on a few broad classes of photonic technologies which have not yet been used in astronomy.

11.1.1 *Microstructured fibres*

All the devices discussed throughout Part II use step-index waveguides which work by total internal reflection. However, graded index fibres (Section 5.2) and microstructured fibres, such as photonic crystal fibres (Section 5.6) and photonic bandgap fibres (Section 5.6.2) offer behaviours not possible with standard step-index fibres. We will look at some potential applications of these devices now.

11.1.1.1 *Single-mode fibre*

We have already mentioned the endlessly single-mode in Section 5.6, which permits single-mode behaviour over a very wide wavelength range compared to step-index fibre. Such fibre thus has the potential to significantly extend the wavelength range of any instrument requiring single-mode behaviour, such as the spatial filtering SMFs for interferometry (Chapter 7) or single-mode spectrographs (Chapter 9). Furthermore, these can be made with a large range of core diameters, and consequently a large range of mode areas, allowing for greater choice in the instrument design. Recall however, that a large mode area *does not* break the requirement to conserve étendue when feeding the fibre (see Chapter 6).

11.1.1.2 *Orbital angular momentum*

Photons carry linear momentum, as is well known from the quantum mechanical relationship $p = \frac{E}{c} = \frac{h}{\lambda}$. Photons also carry a spin angular momentum, due to the quantum mechanical spin of a photon, which is always $\pm\hbar$, and is directly associated with the circular polarisation of the light.

Recently, there has been much attention given to the **orbital angular momentum** of light, which is *not* associated with polarisation, but rather when the wavefront of the light is helical, see e.g. Figure 11.1.

The orbital angular momentum of light from astronomical sources has not been well studied. In part this is because it is a very difficult measurement to make, since atmospheric turbulence (Section 2.1) erases the incoming OAM, and although it has been shown that AO correction is capable of preserving OAM information (Neo *et al.* 2016), the technique has seen little application. Nevertheless, there are several good reasons to make astronomical OAM measurements (Harwit 2003), perhaps the most compelling of which is the detection of the orbital angular momentum from a spinning black hole, from radiation scattered off the black hole.

Recent experiments with twisted multi-core PCF, in which a PCF with several cores is twisted, as sketched in Figure 11.2, have demonstrated the ability to preserve both the magnitude and the chirality of OAM (Beravat *et al.* 2015). These fibres could be used to enable multi-object OAM instruments.

11.1.2 *Phase control*

We have seen a number of astrophotonic components that make use of the phase of the light travelling along a waveguide, e.g. FBGs (Chapter 8) and ABCD interferometry (Chapter 7). However, so far none of these actively controls the phase of the light. The ABCD combiner does shift the phase of the incoming light, but in a fixed way. We now turn our attention to methods and uses of arbitrarily shifting the phase.

In Chapter 6, we discussed the necessity of using multimode fibre for efficient coupling from a telescope in seeing limited conditions. This problem can be alleviated using adaptive optics to correct the wavefront before injecting the light into a fibre, and in the ideal diffraction limited case up to 87% of light could be coupled from a telescope into a SMF in the image plane. We now discuss a photonic solution to this problem, which in principle could enable even higher injection into SMF from a telescope.

11.1.2.1 *Photonic adaptive optics for single-mode spectroscopy*

Consider the device sketched in Figure 11.3 consisting of nine parallel waveguides. If light is injected into the central waveguide alone, then as the light propagates along the waveguide it will evanescently couple

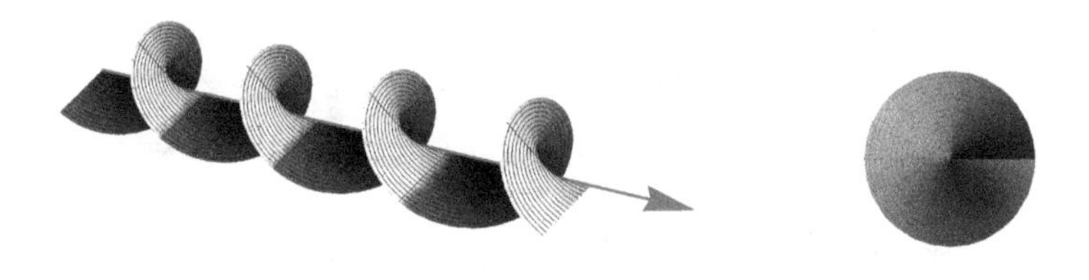

(a) m = 1

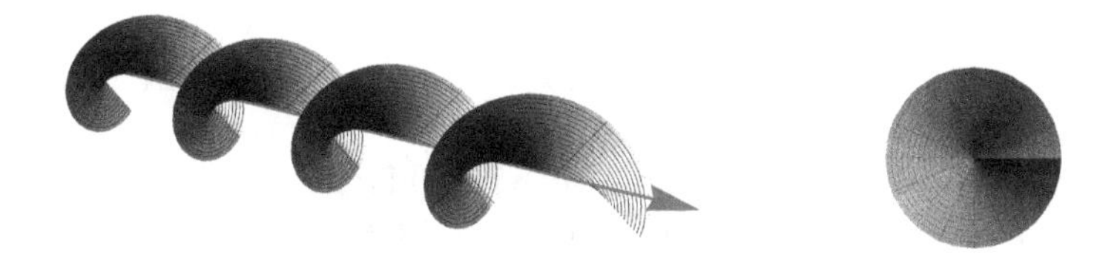

(b) m = −1

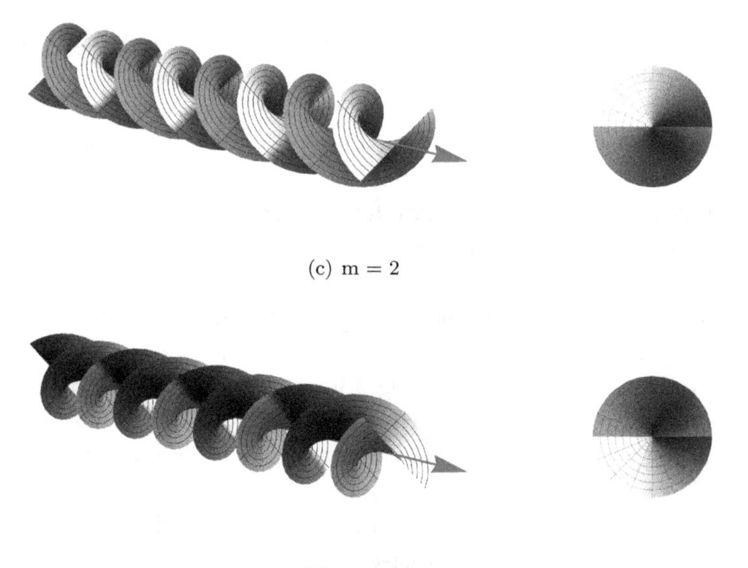

(c) m = 2

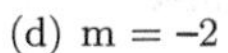

(d) m = −2

Fig. 11.1. Illustration of helical wavefronts giving rise to orbital angular momentum. The phase of the wavefront is depicted by the colour, as shown on the wavefronts, and for a single period on the right hand plots.

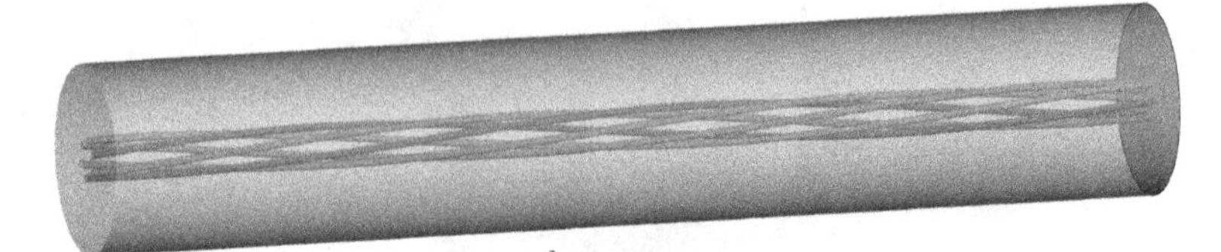

Fig. 11.2. Sketch of a twisted PCF, which preserves both the sign and magnitude of orbital angular momentum, since different chiralities rotate in opposite directions as the light propagates down the fibre. Only the inner holes are shown, and are differently coloured, for clarity.

(Section 6.5.2) into the adjacent waveguides, as studied by Somekh *et al.* (1973).

The operation of such a device can be understand following the normal mode analysis given in Sections 6.5.2 and 6.6. We will not carry out the analysis here since the equations are very long; the procedure is identical to before.

The results are shown in Figure 11.3 when light is initially injected into the central core only. Initially the light starts to propagate to the outer waveguides, and thereafter it starts to couple back in, leading to a complicated interference structure, which will vary as a function of length and wavelength.

Since Maxwell's equations are symmetric with time, this process must work in reverse. If light is injected into the nine waveguides with exactly the correct amplitude and phase for the given wavelength and length of waveguide, then the light will evanescently couple ending up in central single-mode waveguide.

This idea leads to the concept sketched in Figure 11.4. The basic idea is to use a photonic lantern (Section 6.3.2) to feed the channel waveguides. The outputs of all the outer waveguides are constantly monitored, and the appropriate phase modulation is applied to the individual waveguides to ensure that all light is evanescently coupled to the central waveguide only. There are several ways in which the phase of light in a waveguide can be modulated. For example, the waveguide could be physically stretched or heated, a phase change material could be embedded into the waveguide, which can be controlled using a heater, or a Mach–Zehnder waveguide interferometer, such as that sketched in Figure 11.5 could be used, again with the modulation controlled by a heater. We now outline two approaches to applying this scheme.

(a) Length = 30

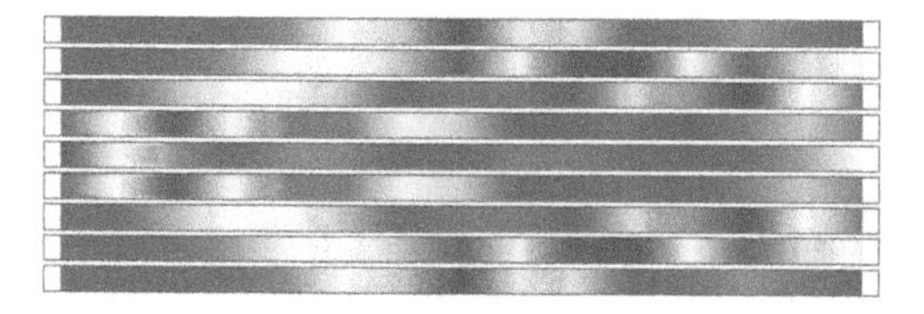

(b) Length = 60

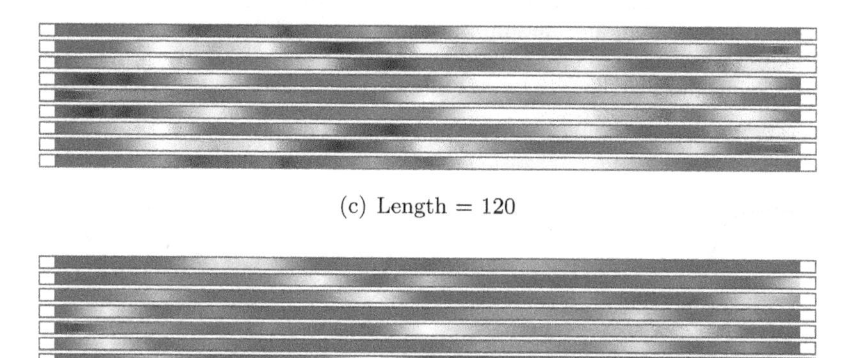

(c) Length = 120

(d) Length = 120

Fig. 11.3. The coupling between nine parallel waveguides for which the light is injected into the central core. The figures are for different lengths of 30, 60, 120 and 120 in arbitrary units. (a)–(c) The intensity scaling is adjusted independently for each waveguide, and (d) all the waveguides are drawn with the same intensity scale.

The first possibility is to calculate the modes in each waveguide (using the same procedure as in Section 6.6) to yield the equivalent relationships as Equations 6.56 and 6.59, but for an arbitrary combination of inputs. The measured power in each waveguide will yield the relative amplitudes at input, which can then be corrected by applying an appropriate phase shift to each waveguide to yield the correct combination of inputs to allow coupling into the single centre waveguide. The monitoring and phase shifting

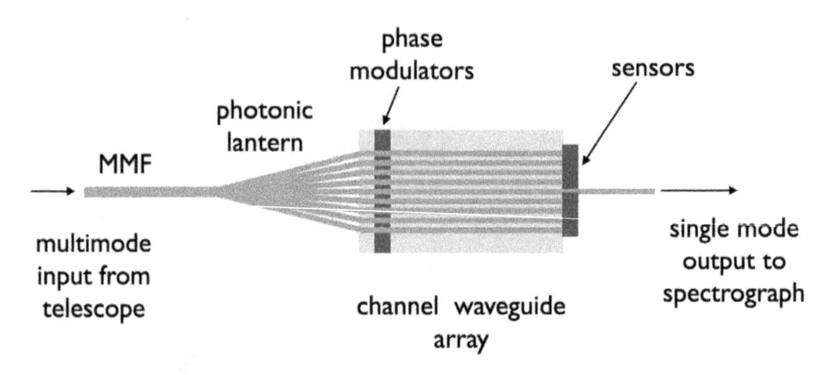

Fig. 11.4. Sketch of the concept for photonic adaptive optics. Light from the telescope is collected by a multimode fibre before being converted into an array of single-mode fibres with a photonic lantern. These SMFs feed a parallel array of channel waveguides. The outputs of all waveguides except the central one are constantly monitored, and an appropriate phase modulation is applied to each individual waveguide to ensure that the output of the central waveguide is maximised, while those of the outer waveguides are zero; see the discussion in the text.

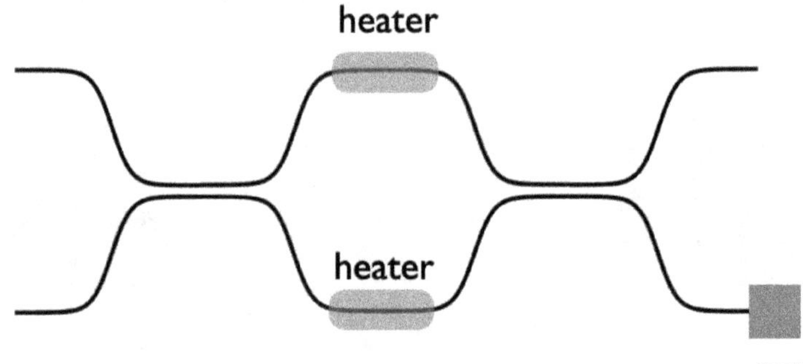

Fig. 11.5. Illustration of a Mach–Zehnder interferometer in an integrated photonic circuit. There are two input channels and two output channels. The phase of light is controlled at the coupling region with a heater. In this way the resulting interference can be made to split the light into any desired fraction at the output waveguides.

needs to be made in real time to compensate for the changing wavefront. Indeed, recently Norris *et al.* (2020) have demonstrated a photonic lantern wavefront sensor (WFS) based on this concept, in which the outputs of the photonic lantern can be used to determine the input wavefront.

The second technique, as explored by Miller (2013), would divide the problem up, by working on pairs of waveguides at a time, and adjusting

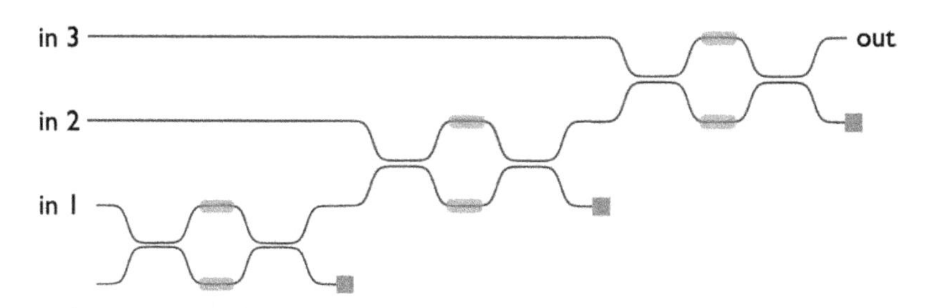

Fig. 11.6. Example of using an iterative approach to provide injection into a single-mode waveguide, based on a series of cascaded Mach–Zehnder interferometers, which are independently adjusted.

the phase of until all light is couple into a single output, and repeating the process until all the light is in a single output. This scheme is sketched in Figure 11.6.

Note that this scheme of 'photonic AO' injection is not really adaptive optics in the traditional sense, since it is impossible to form images in this way. However, the filtering of spatial modes in this way, while still capturing a large fraction of the input light, will provide a very stable PSF for spectroscopy, as discussed in Chapter 9.

The chief difficulty with realising such a technique will be to make it work over a sufficiently broad wavelength range to be useful for spectroscopy. The propagation constants of the waveguides is a function of wavelength (Equation 4.20), and so the phase shifts necessary to couple light into a single output will also depend on wavelength.

11.1.2.2 *Coronagraphy*

Controlling the phase of incoming light can also be used for coronagraphy. Classical coronagraphy uses a physical mask in the image plane to block out the light from the Sun, and hence allow observations of its corona. The same technique has also been used to block out light from stars to search for exoplanets.

A novel extension of this idea is to replace the physical mask with a phase mask which imparts an azimuthal phase change on the incoming light (as in the right hand panels of Figure 11.1). This diffracts light from the centre of the image to large angles which can then be masked using a pupil stop, before refocussing the light into an image without the central star (Mawet *et al.* 2009, 2010).

Similarly, phase-induced amplitude apodisation can impart a phase change at an intermediate pupil to provide apodisation, i.e. to soften the edges of the pupil. This has the effect of removing the Airy rings in the diffraction limited point spread function (PSF), which otherwise hamper efforts to detect the faint light from exoplanets (Guyon *et al.* 2005).

More recently, Fontaine *et al.* (2019) have demonstrated a method to decompose an image into separate Laguerre–Gaussian modes using a spatial light modulator. Selectively removing any on-axis ($l = 0$) modes before recomposing the image can provide another means of rejecting light from a particular point on the image plane (Carpenter *et al.* 2020).

11.1.3 *Polarisation*

Many devices presented in this work rely on channel waveguides with rectangular cross-sections. As already noted these are intrinsically birefringent with different propagation constants for TE and TM modes (Section 4.2.3). Although in some cases this is a nuisance, since often we want to guide and manipulate all the captured light irrespective of its polarisation, in cases where we are specifically interested in the polarisation of the light this can be an advantage.

There are many types of polarisation splitters which can be incorporated within integrated optics circuits, for example, based on the geometry of the waveguides, optical fibres with asymmetries in the cores, asymmetric Mach–Zehnder interferometers, or highly birefringent polymers (see, e.g. Huang *et al.* 2017 and references therein). These devices have been demonstrated in various materials including, silicon, silica, and polymer.

These polarisation splitters and waveguides offer an immediate application for both photometric polarisation studies, or if combined with arrayed waveguide gratings (or other WDM devices) for spectropolarimetry. Furthermore, since we are by definition dealing with waveguides, this opens up the possibility of multi-object polarisation studies.

CHAPTER 12

THE FUTURE OF ASTROPHOTONICS

In the last chapter, we examined various existing photonic devices, or devices under development, which have not yet been exploited in astrophotonics, but which offer obvious potential.

We now turn our attention to future developments which have not yet begun, or which exist only at the broad concept level. Thus, whilst the previous chapter explored the future of photonics from the point of view of examining devices which have been developed for other applications, but which could in principle be adapted for astronomy, we now look at the problem the other way around, i.e. we examine what developments in photonics are most needed from the point of view of astronomy.

12.1 Multimode Photonics

A recurring theme throughout this book has been the need to use multimode waveguides to collect sufficient light from the telescope focal plane, and the subsequent need to separate the light into single-mode waveguides, e.g. via a photonic lantern, to allow the use of standard photonic devices. Since many photonic components are usually mass producible and modular, this is not necessarily a major disadvantage, as long as the yield fraction of manufactured devices is high and the losses at the interfaces are low.

Nevertheless, it could be possible to avoid the need for highly replicated astrophotonic devices with multimode photonics. In some senses, this idea has been approached in astrophotonics with the use of multicore fibres (Section 6.3.2.1), in which all the separate single-mode cores are contained within the same fibre, which can make certain devices, such as FBGs, more efficient and easier to write (see Bland–Hawthorn *et al.* 2016 for a discussion).

In fact multimode photonics is also desirable for other photonic applications, for example mode division multiplexing, in which separate signals can be multiplexed and demultiplexed in different modes of the same waveguide. In this case, it is important that the cross-talk between modes in the waveguide is very small so that the signals do not become mixed. In the case of astronomy this will not usually be important, since all the modes are carrying the same information, and the large number of modes is necessary only to increase the grasp of the instrument.

Work in this area is only just beginning and is as yet restricted to cases of a few modes; (see, e.g. Li *et al.* 2019; Ramachandran and Parmigiani 2019 for reviews of multimode silicon and fibre photonics). Nevertheless, some promising early results have been obtained, including WDM multimode devices based on Bragg gratings (Xie *et al.* 2018) and ring resonators (Luo *et al.* 2014). So far, all these devices use a combination of single-mode and multimode waveguides to achieve the desired function, see e.g. Figure 12.1 for schematic examples of multimode WDM, but importantly they are manufactured on a single chip, with no requirement for mass replication. This area will very likely become more important following further development in the coming years.

12.2 Fully Photonic Instruments

So far, all the devices discussed in this book must be incorporated as a part of more traditional astronomical instruments, such as spectrographs and cameras. However, the ultimate astrophotonic instrument would require no bulk optics. Light would be coupled into a waveguide, and thereafter all functions would be carried out in photonic devices up to detection of the photons. This is not simply an ideological desire, but would allow real benefit, enabling replication and modularity and minimising losses due to interconnections. Moreover, as many of these functions as possible should be carried out on the same photonic chip.

As an example of this consider a variation on the PIMMS instrument first proposed by Bland–Hawthorn *et al.* (2010), sketched in Figure 12.2. Light is collected at the telescope focal plan by a multimode fibre, which feeds an array of single-mode fibres via a photonic lantern. Each SMF feeds

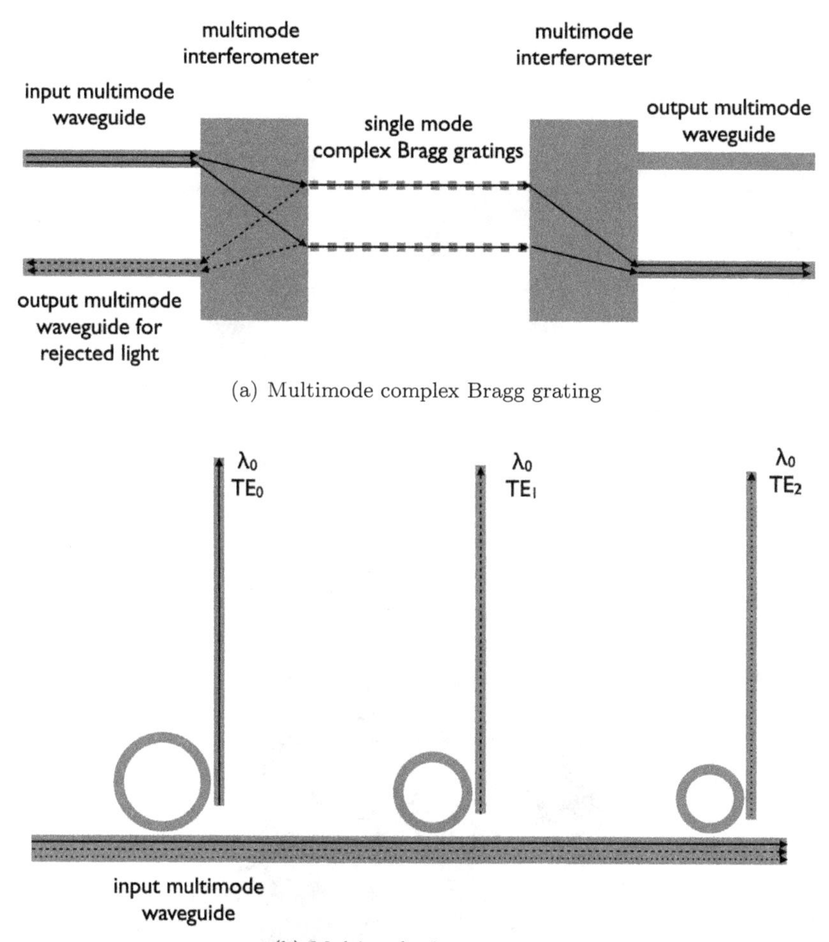

(a) Multimode complex Bragg grating

(b) Multimode ring-resonator

Fig. 12.1. Diagrams illustrating the principals of two multimode WDM schemes. Panel (a) is based on Xie *et al.* (2018), which uses a multimode input and a multimode interferometer to feed each mode to a separate single-mode waveguide in which is written a complex Bragg grating. Symmetrical interferometers on output recombine the two transmitted signals to a multimode output. Xie *et al.* (2018) demonstrated such a device capable of rejecting five separate wavelengths. In principle, the idea can be extended to more modes. In panel (b) is a device following the ideas of Luo *et al.* (2014), in which an input waveguide carries several modes. Each mode is filtered by an individual ring resonator, each of which is tuned to the same wavelength, but which only couples to one of the modes.

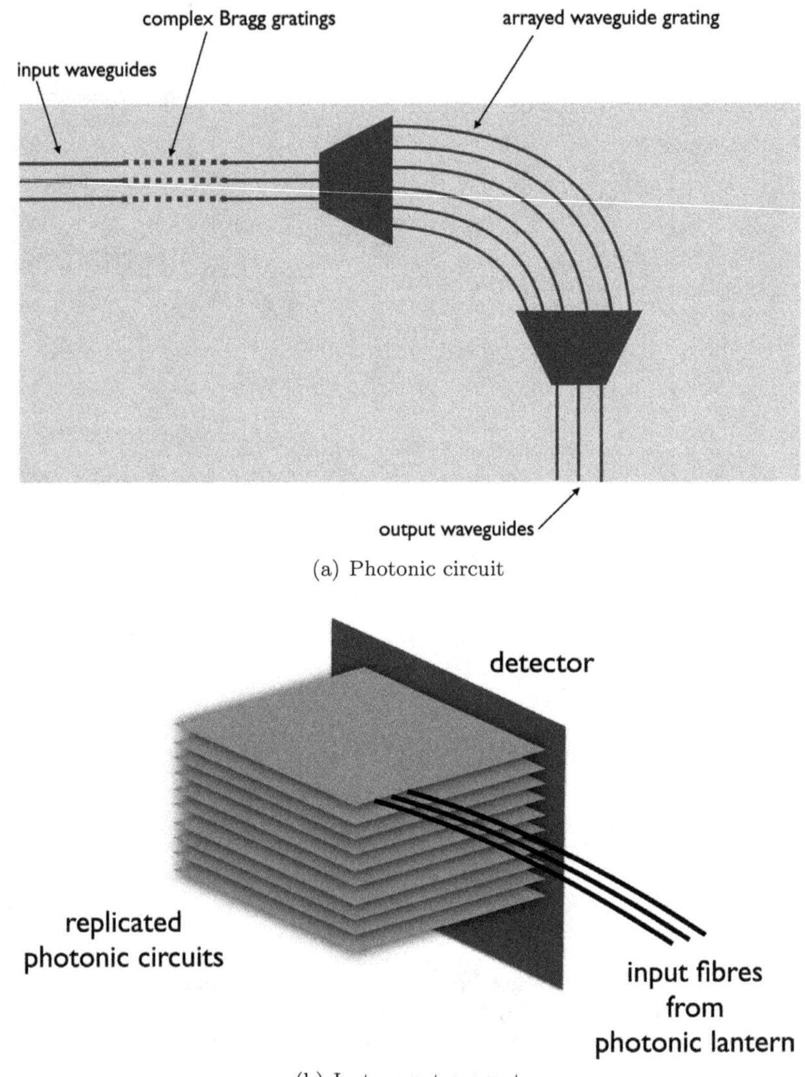

(a) Photonic circuit

(b) Instrument concept

Fig. 12.2. Sketch of a concept for a fully photonic instrument following Bland–Hawthorn *et al.* (2010), combining complex Bragg waveguide gratings and arrayed waveguide gratings. Only three waveguides are shown for clarity.

a single-mode waveguide on a photonic circuit. The waveguides are first split into orthogonal polarisations, such that subsequent steps can be carried out with devices tuned to the effective index of each polarisation. Each waveguide then feeds a complex Bragg grating (or a series of ring resonators) to reject the atmospheric OH lines. From here each waveguide can feed an arrayed waveguide grating (or a photonic echelle grating or Lippmann spectrograph) on the same chip, providing spectroscopic capability. The light is finally recorded on a butt-coupled detector. The output of many stacked chips can be recorded on the same detector array. Of course, these are not the only particular devices which may combined on the same chip, and we could have instead chosen to illustrate the concept with photonic AO SMF injection, polarisation splitters, interferometric beam combiners, etc.

Whilst certainly futuristic, such an instrument has many compelling features. In the first place it would be tiny compared to an equivalent bulk object instrument carrying out the same functions. Second, it is very modular, and could easily be extended to include more inputs enabling multi-object spectroscopy or IFS, etc.

The final step in the concept sketched in Figure 12.2 is the detection of the photons with a butt-coupled detector. Since 2D detector arrays are ubiquitous in astronomy, and have excellent QE and noise characteristics, such a solution is a natural choice. However, looking to the future it is not necessarily the ideal choice. In the first place, the concept shown would actually require a cross-dispersing element between the waveguide output and the detector in order to separate the overlapping orders from the AWG (Section 9.2.2). This situation could be improved in principle, by using an energy resolving detector such as an MKIDS device (O'Brien 2020).

However, ultimately it is not clear that a 2D detector array is the optimal format for fully photonic instruments, since a 2D array is not a natural fit to a planar circuit. Rather, a 1D array of detectors would be better suited.

One way to miniaturise such instruments even further is to use detectors embedded in the waveguides themselves. It is not our purpose to review such devices here, but we note that several schemes already exist, including photodiodes, super-conducting nanowire single photon detectors (e.g. Höpker *et al.* 2019), and integrated transition edge sensors (e.g. Ferrari *et al.* 2018).

12.3 Quantum Astrophotonics

Throughout this book, we have presented a diverse array of techniques and applications under the banner of astrophotonics. One of the unifying characteristics of all these techniques is the use of waveguides to perform some function. Another distinction, at a perhaps more fundamental level is in the treatment of light itself. As discussed in the introduction, for most classical astronomical instruments it is sufficient to use a geometric or wave-optics approximation in the treatment of light in order to design and understand the instruments. By contrast, photonics often requires a full electromagnetic approach to understand and design the devices, especially regarding the description of modes, and coupling between waveguides.

At a deeper level still, a fuller description of the interaction of light with matter is found in quantum electrodynamics. The field of optics which requires a full quantum approach is known as quantum optics. For the final part of this book, we turn our attention to quantum astrophotonics.

12.3.1 *Quantum optics*

In 1948, Hanbury Brown arrived at an important insight. If a single telescope is pointed at a bright celestial source, one observes that the received signal is fluctuating rapidly as a consequence of the signal's intrinsic noise and extraneous background noise. Two telescopes placed side by side would observe the same source to be rapidly fluctuating, and some part of that fluctuating signal would appear correlated between the two telescopes. As we move the telescopes farther apart, the correlation would slowly fade until the signals appear uncorrelated. Hanbury Brown argued that the spatial extent (projected onto the sky) of the observed correlation would be directly related to the angular size of the celestial source — a technique known as intensity interferometry.

After a successful trial in the UK at radio wavelengths, Hanbury Brown and Lord Twiss succeeded in demonstrating the same effect at optical wavelengths in Narrabri, Australia. The idea that a star's fluctuating intensity is correlated at the photon level between two separated telescopes was not universally believed, even after they demonstrated the same phenomenon in the lab. But after Purcell (1956) provided an explanation that unified both classical wave theory and a quantum description, the field of quantum optics was born. We now speak of 'HBT interferometry' to describe two-photon correlation used in intensity interferometry.

While the birth of quantum optics emerged from astronomy, it has had limited application on telescopes since about 1974. Hanbury Brown is reputed to have said that his technique failed to find important uses in astronomy, but history has been kinder to him. The angular diameters of 32 hot stars measured by his team ultimately led to the first reliable measurements of surface fluxes and temperatures, and these formed the bedrock of stellar flux calibrations. HBT interferometry is also used extensively in nuclear, heavy ion and ultra-cold atomic physics (Kleppner 2008). After the last HBT papers appeared in the early 1970s, it was not revisited by astronomers until Guerin *et al.* (2018); this reference lists eight groups worldwide engaged in related activities. With the aid of new enabling technologies such as photonics and fast-cadence detectors, we are likely to see more extensive use of intensity interferometry in the extremely large telescope (ELT) era. We expand on this statement in the following section.

12.3.2 *Quantum photonics*

In the present decade, there is heavy investment in the idea of quantum computing, where superposition and entanglement are used to perform computations. There are now hundreds of research groups invested in this development, and some have foreseen the need for a quantum network to allow the quantum computers to share information. In order to achieve a quantum internet, widely separated quantum processors must be able to send quantum information (qubits) reliably from node to node, a task known as quantum teleportation. The notion of quantum encryption, key distribution (QKD) and communications is one of the fastest-developing fields within this sphere. The drive to develop quantum networks has led to the emergence of a new field — quantum photonics — that underpins the connectivity of these components, much as conventional photonics has enabled telecomm. networks. Recent and anticipated developments in quantum photonics paves the way for a new generation of astronomical telescopes, as we describe.

Embedded within the idea of a quantum internet is the prospect of collecting photons, more specifically quantum states, at widely separated locations over the Earth's surface. This is the first step to achieving optical or infrared interferometry over Earth-diameter baselines (optical VLBI; Bland–Hawthorn *et al.* 2021). The question is how to bring these probability amplitudes together in order to generate an interference pattern, for example. Quantum entanglement experiments using polarised photons maintain their coherence over a baseline of about 100 km. Since the best single-mode communications fibre has 0.2 dB/km insertion losses, this means that 99% of the photons are lost along its length. Free-space propagation, say from a passing satellite, are orders of magnitude more lossy. Since you cannot make multiple identical copies of an unknown quantum state without altering the original state — i.e. there is no quantum cloning — a new approach is called for.[1]

To underscore the fundamental achievement of propagating a delicate quantum state, consider this statement from quantum information theory. The **no-teleportation theorem** states that an arbitrary quantum state cannot be converted into a (finite or infinite) sequence of classical bits, nor can such bits be used to reconstruct the original state. Thus, teleporting a state by merely moving classical bits around does not work because the unit of quantum information, the qubit, cannot be precisely converted into classical information bits because of the Heisenberg uncertainty principle.

This should not be confused with quantum teleportation in Figure 12.3 that does allow a quantum state to be destroyed in one location, and an

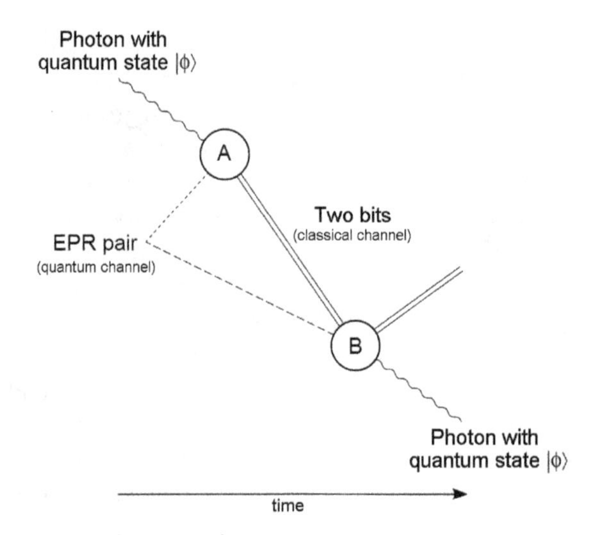

Fig. 12.3. Schematic drawing of a quantum repeater, one segment of a quantum network. A future quantum network is envisaged to be made up of many of these segments with the aid of quantum memories placed at each node. Courtesy of Wikipedia, Public Domain.

[1] Interestingly, imperfect copies can be made, a concept known as quantum copying (Bužek and Hillery 1996); this concept is potentially useful in quantum interferometry with regard to maintaining the metrology and stability of the system.

exact replica to be created at a different location. A promising technology to achieve this is the quantum repeater (Figure 12.3) that has only been demonstrated in the lab to date.

Consider how it might be possible to interfere 'science' photons arriving at each node, A and B. The exact quantum state of a photon $|\phi_A\rangle$ is not preserved as it propagates in a conventional telecomm. fibre. In the quantum repeater, a pair of entangled photons is generated at a trusted site, and one of each is transmitted to nodes A and B over a polarisation-preserving communication (fibre) channel. The idea here is that the entanglement is preserved through their relative polarisations, as expected from a biphoton source using nonlinear parametric down conversion. The entangled photon arriving at A is mixed with the science photon through a 'Bell test circuit' that in turn delivers 2 qubits on output. This information is transmitted over a regular telecomm. cable from A to B. At node B, the 2 cubits are mixed with the entangled photon arriving at B using another Bell measurement. It can be shown that the resulting photon from the Bell circuit (reverse operation relative to the first Bell test) is the original quantum state $|\phi_A\rangle$. The transmitted state over the quantum repeater preserves all information including amplitude, phase and higher-order (e.g. orbital angular momentum) states. We are now at liberty to interfere the science photon from node A, $|\phi_A\rangle$, with a science photon harvested locally at node B, $|\phi_B\rangle$.

In Figure 12.3, a time-ordered sequence of events is indicated. If we were to build a quantum network from the quantum repeater segments, this would manifestly fail if we were not able to store complete quantum states at the different nodes before the information is made available for transmission to another node. Over a distance of 100 km, the light travel time τ is of order a microsecond. The synchronicity issue across the network requires a quantum memory (QM) to retain its information for longer than τ. Quantum memories that store their information long enough to achieve coherent entanglement between two distant nodes has become a reality, with a recent QM–QM entanglement achieved over a distance of 22 km (Yu *et al.* 2020). Some memories have managed to preserve an exact quantum state for up to six hours but not yet within a repeater (Zhong *et al.* 2015). This is a rapidly developing field.

Conventional computers distinguish between random access memory (temporary storage) and the hard drive (long-term storage). Along the same lines, one can imagine a 'quantum hard drive' (QHD) that is able to store quantum information for days or even years. Such devices are already under development. Thus, if we were able to store quantum information at node A on a QHD, this device could then be transported physically to node B. The stored signals on a local QHD would then be 'read' and combined with the other QHD, or QHDs if this is a multi-node network. The power of this idea comes from realising that the longest baselines for optical interferometry are about 100 m, whereas QHDs could conceivably allow us to combine signals from optical/infrared telescopes dispersed across the globe — in essence, optical VLBI that would allow us to extend beyond microarcsecond to nanoarcsecond astrometry. This has many implications, inter alia, all stars in the Galaxy, are now potentially resolvable, assuming there are sufficient photons from the targeted source. In an era of ELTs, such a notion is more than theoretical.

12.3.3 *Quantum future*

Looking to more distant horizons, we can conceive of a future when the properties of optical and infrared photons are tagged individually in time and in space. High-energy astrophysics treats individual events in this way, in particular, cosmic rays, neutrinos, hard X-ray and gamma-ray photons. These are often rare events emerging from explosive astrophysical sources (e.g. supernovae, magnetars, microquasars). But here, we are imagining vastly higher data rates from low-energy (optical, IR) photons more typical of observational astronomy, often limited by background radiation.

Consider the prospect of storing a complete quantum description of billions of detected photon 'events' in a giant quantum memory. These 'events' could be recovered at will and post-processed in order to make an (astronomical) observation. Rather than exploiting conventional optics to achieve an observation, we foresee a quantum computer will be used to manipulate the stored data. If the requested observation is an analogue interferogram, the necessary delay will be imposed as a function of the photon's position on the sky and the spacing of the receiver from the other receivers. If the requested observation is an analogue spectrum, the necessary delay will be imposed as a function of the photon's energy.

But there are fundamental rules at play. It is tempting to think that the 'event' in storage can be read many times to increase the signal-to-noise ratio of the 'event' but this is not so. Quantum cloning is not allowed, such that when the event is read, it is lost to the memory. Some of the technologies needed to make this a reality are already under development. While still a long way off, the worldwide development of quantum computers and quantum networks is leading to innovations that are likely to find uses in other fields. We foresee a future for astrophotonics where some of these innovations are adapted for astronomy (Bland–Hawthorn *et al.* 2021).

12.4 The Future of Astrophotonics

Throughout this book, we have seen the remarkable potential of astrophotonics for astronomical instrumentation. In Part II, we have illustrated the breadth of photonic technologies currently in use or under development, which already cover a very wide range of applications. Indeed, some of these astrophotonic instruments have already had a transformative impact on astrophysics, most especially the gains resulting from photonic interferometric beam combiners, and several others are on the cusp of making a similar impact, e.g. OH suppression. Thus, astrophotonics is currently in a period of transition, as many instruments are progressing from concepts to prototypes to full facility class instruments, and the pace of development is likely to increase as a greater number of photonic technologies are adapted for astronomy.

In the final part of the book, we have suggested ways in astrophotonics may develop in the future. In Chapter 11, we looked at existing photonic technologies, which have not been exploited, or which have received only little attention, but which have the possibility to make a significant impact. In this chapter, we have considered future developments based on ideas at the forefront of current research, and motivated by the requirements of astrophysics. These prospective developments can be expected to increase the range of astrophotonics even further.

Winston Churchill once commented that we must be wary of needless innovation, especially when guided by logic. This is particularly true of 'big science' where today's high-profile physics experiments (e.g. Large Hadron Collider, LIGO) will continue to scale up because this has been the way of things for centuries. The largest telescope at the close of the 18th century was Herschel's 1.2 m telescope, followed by Lord Rosse's 1.8 m telescope in the century that followed. The 5.1 m Hale telescope dominated astronomy until the rise of 8–10 m class telescopes in the 1990s. Today, we stand at the dawn of a new era of (25–40 m) ELTs. It is conceivable that this scaling-up will continue until there is a breakthrough in the underlying technology or a fundamental limit is reached. These telescopes will have such enormous light-gathering potential that the time has come to consider new ways of doing astronomy.

Astrophotonics will play a key role in this development, offering radical new ways to manipulate light, far beyond the traditional iterations and scaling-up of previous instruments. Indeed, astrophotonics is especially apposite to the ELT era. Adaptive optics and near-diffraction limited performance are designed into the telescopes at the outset, with the effect that coupling into astrophotonic devices will be much simpler and more efficient. Moreover, the ease of replication and the modularity of photonic components allows a new approach to instrumentation which can end the perpetual increase of instrument size, and telescopes get bigger. That is to say, ELTs will make astrophotonics easier, and in return astrophotonics will make ELTs more productive and powerful.

It has been said that the best way to predict the future is to invent it. We hope that with this monograph we have provided the tools with which future instrument scientists, astronomers and photonicists can understand the bases and principles which underpin this exciting field, and that someone reading this book takes it to heart and makes the first step in inventing the instruments of the future.

BIBLIOGRAPHY

Abrams, M. C., Davis, S. P., Rao, M. L. P., *et al.* (1994). High-resolution Fourier transform spectroscopy of the Meinel system of OH, *ApJS* **93**, pp. 351–395, doi:10.1086/192058.

Allington-Smith, J. and Bland-Hawthorn, J. (2010). Astrophotonic spectroscopy: Defining the potential advantage, *MNRAS* **404**, 1, pp. 232–238, doi:10.1111/j.1365-2966.2009.16173.x, arXiv:0910.4361 [astro-ph.IM].

Ams, M., Marshall, G. D., Dekker, P., *et al.* (2008). Investigation of ultra-fast laser-photonic material interactions: Challenges for directly written glass photonics, *IEEE Journal of Selected Topics in Quantum Electronics* **14**, 5, pp. 1370–1381, doi:10.1109/JSTQE.2008.925809, arXiv:0802.1966 [physics.optics].

Bacon, R. and Monnet, G. (2017). *Optical 3D-Spectroscopy for Astronomy* (Wiley-VCH, Weinheim).

Bailey, J., Cotton, D. V., Kedziora-Chudczer, L., *et al.* (2020). HIPPI-2: A versatile high-precision polarimeter, *PASA* **37**, e004, doi:10.1017/pasa.2019.45, arXiv:1911.02123 [astro-ph.IM].

Barden, S. C. (1987). *KPNO Fiber Optics Lab. Report 1* (Kitt Peak National Observatory, Tucson).

Bely, P. Y. (2003). *The Design and Construction of Large Optical Telescopes* (Springer, New York).

Beravat, G. K. L., Wong, R., Xi, X. M., *et al.* (2015). Preservation of magnitude and chirality of oam order in continuously twisted pcf with six satellite cores, *CLEO/Europe-EQEC*, p. CI3.3.

Berger, J. P., Haguenauer, P., Kern, P., *et al.* (2001). Integrated optics for astronomical interferometry. IV. First measurements of stars, *A&A* **376**, pp. L31–L34, doi:10.1051/0004-6361:20011035.

Betters, C. H., Leon-Saval, S. G., Bland-Hawthorn, J., *et al.* (2012). Demonstration and design of a compact diffraction limited spectrograph, *Proceedings of SPIE* **8446**, p. 84463H.

Betters, C. H., Leon-Saval, S. G., Robertson, J. G., *et al.* (2013). Beating the classical limit: A diffraction-limited spectrograph for an arbitrary input beam, *Optics Express* **21**, 22, p. 26103, doi:10.1364/OE.21.026103, arXiv:1310.4833 [astro-ph.IM].

Betters, C. H., Leon-Saval, S. G., Bland-Hawthorn, J., *et al.* (2014). PIMMS échelle: The next generation of compact diffraction limited spectrographs for arbitrary input beams, *Proceedings of SPIE* **9147**, p. 91471I.

Bharathan, G., Fernandez, T. T., Ams, M., *et al.* (2019). Optimized laser-written ZBLAN fiber Bragg gratings with high reflectivity and low loss, *Optics Letters* **44**, 2, p. 423, doi:10.1364/OL.44.000423.

Birks, T. A., Knight, J. C., and St. Russell, J. P. (1997). Endlessly single-mode photonic crystal fiber, *Optics Letters* **22**, p. 961.

Birks, T. A., Gris-Sánchez, I., Yerolatsitis, S., *et al.* (2015). The Photonic Lantern, *Advances in Optics and Photonics* In press, arXiv:1503.02837 [physics.optics].

Blais-Ouellette, S. (2004). Holographic gratings for astronomy: Atmospheric lines suppression and tunable filter, in *Photonics North 2004: Photonic Applications in Astronomy, Biomedicine, Imaging, Materials Processing, and Education*, Vol. 5578, pp. 23–28, doi:10.1117/12.567440.

Blais-Ouellette, S., Artigau, É., Havermeyer, F., *et al.* (2004). Multi-notch holographic filters for atmospheric lines suppression, in *Optical Fabrication, Metrology, and Material Advancements for Telescopes*, Vol. 5494, pp. 554–561, doi:10.1117/12.552116.

Bland-Hawthorn, J. and Horton, A. (2006). Instruments without optics: An integrated photonic spectrograph, *Proceedings of SPIE* **6269**, p. 62690N.

Bland-Hawthorn, J., Buryak, A., and Kolossovski, K. (2008). Optimization algorithm for ultrabroadband multichannel aperiodic fiber Bragg grating filters, *Journal of the Optical Society of America A* **25**, p. 153.

Bland-Hawthorn, J., Englund, M., and Edvell, G. (2004). New approach to atmospheric oh suppression using an aperiodic fibre Bragg grating, *Optics Express* **12**, p. 5902.

Bland-Hawthorn, J., Ellis, S., Haynes, R., *et al.* (2009). Photonic OH suppression of the infrared night sky: First on-sky results, *Anglo-Australian Observatory Newsletter* **115**, p. 15.

Bland-Hawthorn, J., Kos, J., Betters, C. H., *et al.* (2017). Mapping the aberrations of a wide-field spectrograph using a photonic comb, *Optics Express* **25**, 14, p. 15614, doi:10.1364/OE.25.015614, arXiv:1704.08775 [astro-ph.IM].

Bland–Hawthorn, J., Min, S.-S., Lindley, E., *et al.* (2016). Multicore fibre technology: The road to multimode photonics, in *Advances in Optical and Mechanical Technologies for Telescopes and Instrumentation II*, Vol. 9912, p. 99121O, doi:10.1117/12.2231924.

Bland–Hawthorn, J., Lawrence, J., Robertson, G., *et al.* (2010). PIMMS: Photonic integrated multimode microspectrograph, in *Society of Photo-Optical Instrumentation Engineers (SPIE) Conference Series, Society of Photo-Optical Instrumentation Engineers (SPIE) Conference Series*, Vol. 7735, doi:10.1117/12.856347.

Bland-Hawthorn, J., Sellars, M. J., and Bartholomew, J. G. (2021). Quantum memories and the double-slit experiment: Implications for astronomical interferometry, *JOSA B* **38**, 7, p. A86, doi:10.1364/JOSAB.424651, arXiv:2103.07590 [astro-ph.IM].

Blind, N., Le coarer, E., Kern, P., *et al.* (2017). Spectrographs for astrophotonics, *Optics Express* **25**, p. 27341, doi:10.1364/OE.25.027341.

Bryant, J. J., Bland-Hawthorn, J., Fogarty, L. M. R., *et al.* (2014). Focal ratio degradation in lightly fused hexabundles, *MNRAS* **438**, pp. 869–877, doi:10.1093/mnras/stt2254, arXiv:1311.6865 [astro-ph.IM].

Bryant, J. J., Thomson, R. R., and Withford, M. J. (2017). Focus issue introduction: Recent advances in astrophotonics, *Optics Express* **25**, 17, p. 19966, doi:10.1364/OE.25.019966.

Bužek, V. and Hillery, M. (1996). Quantum copying: Beyond the no-cloning theorem, *Physical Review A* **54**, 3, pp. 1844–1852, doi:10.1103/PhysRevA.54.1844, arXiv:quant-ph/9607018 [quant-ph].

Cai, C. and Wang, J. (2022). Femtosecond laser-fabricated photonic chips for optical communications: A review, *Micromachines* **13**, p. 630.

Carpenter, J., Fontaine, N. K., Norris, B. R. M., *et al.* (2020). Spatial mode sorter coronagraphs, in *14th Pacific Rim Conference on Lasers and Electro-Optics (CLEO PR 2020)* (Optical Society of America), paper C6G_3.

Carrasco, E. and Parry, I. R. (1994). A method for determining the focal ratio degradation of optical fibres for astronomy, *MNRAS* **271**, doi:10.1093/mnras/271.1.1.

Cecil, G. N., Moffett, A. J., Cui, Y., *et al.* (2010). Deployable integral field units, multislits, and image slicer for the Goodman Imaging Spectrograph on the SOAR Telescope, in *American Astronomical Society Meeting Abstracts #215, American Astronomical Society Meeting Abstracts*, Vol. 215, p. 441.

Chromey, F. R. (2010). *To Measure the Sky* (Cambridge University Press).

Connes, P. and Le Coarer, E. (1995). 3D Spectroscopy: The historical and logical viewpoints, in G. Comte and M. Marcelin (eds.), *IAU Colloq. 149: Tridimensional Optical Spectroscopic Methods in Astrophysics*, Vol. 71, p. 38.

Content, R. (1996). Deep-sky infrared imaging by reduction of the background light. I. Sources of the Background and potential suppression of the OH Emission, *Astrophysics Journal* **464**, p. 412.

Corbett, J. C. (2009). Sampling of the telescope image plane using single- and few-mode fibre arrays, *Optics Express* **17**, pp. 1885–1901, doi: 10.1364/OE.17.001885.

Coudé du Foresto, V. and Ridgway, S. T. (1992). Fluor — A stellar interferometer using single-mode fibers, in *European Southern Observatory Conference and Workshop Proceedings*, Vol. 39, p. 731.

Coudé du Foresto, V., Ridgway, S., and Mariotti, J.-M. (1997). Deriving object visibilities from interferograms obtained with a fiber stellar interferometer, *A&AS* **121**, pp. 379–392, doi:10.1051/aas:1997290.

Cvetojevic, N., Lawrence, J. S., Ellis, S. C., *et al.* (2009). Characterization and on-sky demonstration of an integrated photonic spectrograph for astronomy, *Optics Express* **17**, pp. 18643–18650, arXiv:0910.4804 [astro-ph.IM].

Cvetojevic, N., Jovanovic, N., Betters, C., *et al.* (2012a). First starlight spectrum captured using an integrated photonic micro-spectrograph, *A&A* **544**, L1, doi:10.1051/0004-6361/201219116, arXiv:1208.4418 [astro-ph.IM].

Cvetojevic, N., Jovanovic, N., Lawrence, J., *et al.* (2012b). Developing arrayed waveguide grating spectrographs for multi-object astronomical spectroscopy, *Optics Express* **20**, p. 2062, doi:10.1364/OE.20.002062, arXiv:1201.4616 [astro-ph.IM].

Cvetojevic, N., Jovanovic, N., Gross, S., *et al.* (2017). Modal noise in an integrated photonic lantern fed diffraction-limited spectrograph, *Optics Express* **25**, p. 25546, doi:10.1364/OE.25.025546.

Daendliker, R. (2000). Concept of modes in optics and photonics, in J. J. Sanchez-Mondragon (ed.), *Sixth International Conference on Education and Training in Optics and Photonics, Society of Photo-Optical Instrumentation Engineers (SPIE) Conference Series*, Vol. 3831, pp. 193–198, doi:10.1117/12.388718.

Dalton, G., Trager, S., Abrams, D. C., *et al.* (2018). Construction progress of WEAVE: The next generation wide-field spectroscopy facility for the William Herschel Telescope, *Proceedings of SPIE* **10702**, p. 107021B.

Davis, K. M., Miura, K., Sugimoto, N., *et al.* (1996). Writing waveguides in glass with a femtosecond laser, *Optics Letters* **21**, pp. 1729–1731, doi:10.1364/OL.21.001729.

Dinkelaker, A. N., Rahman, A., Bland-Hawthorn, J., *et al.* (2021a). Astrophotonics: Introduction to the feature issue, *Journal of the Optical Society of America B Optical Physics* **38**, 7, p. AP1, doi:10.1364/JOSAB.434565.

Dinkelaker, A. N., Rahman, A., Bland-Hawthorn, J., *et al.* (2021b). Astrophotonics: Introduction to the feature issue, *Applied Optics* **60**, 19, p. AP1, doi:10.1364/AO.434555.

Douglass, G., Dreisow, F., Gross, S., *et al.* (2018). Femtosecond laser written arrayed waveguide gratings with integrated photonic lanterns, *Optics Express* **26**, 2, pp. 1497–1505, doi:10.1364/OE.26.001497, http://www.opticsexpress.org/abstract.cfm?URI=oe-26-2-1497.

Ellis, S. C. and Bland-Hawthorn, J. (2008). The case for oh suppression at near-infrared wavelengths, *MNRAS* **386**, p. 47.

Ellis, S. C., Bland-Hawthorn, J., and Leon-Saval, S. G. (2021). General coupling efficiency for fiber-fed astronomical instruments, *Journal of the Optical Society of America B* **38**, 7, pp. A64–A74, 10.1364/JOSAB.423905, http://josab.osa.org/abstract.cfm?URI=josab-38-7-A64.

Ellis, S. C., Bland-Hawthorn, J., Lawrence, J., *et al.* (2012a). Suppression of the near-infrared OH night-sky lines with fibre Bragg gratings — First results, *MNRAS* **425**, pp. 1682–1695, doi:10.1111/j.1365-2966.2012.21602.x, arXiv:1206.6551 [astro-ph.IM].

Ellis, S. C., Bland-Hawthorn, J., Lawrence, J. S., *et al.* (2020). First demonstration of OH suppression in a high-efficiency near-infrared spectrograph, *MNRAS* **492**, 2, pp. 2796–2806, doi:10.1093/mnras/staa028, arXiv:2001.04046 [astro-ph.IM].

Ellis, S. C., Crouzier, A., Bland-Hawthorn, J., *et al.* (2012b). Potential applications of ring resonators for astronomical instrumentation, in *Society of Photo-Optical Instrumentation Engineers (SPIE) Conference Series, Society of Photo-Optical Instrumentation Engineers (SPIE) Conference Series*, Vol. 8450, p. 1, doi:10.1117/12.925804.

Ellis, S. C., Saunders, W., Betters, C., *et al.* (2014). The problem of scattering in fibre-fed VPH spectrographs and possible solutions, in *Society of Photo-Optical Instrumentation Engineers (SPIE) Conference Series, Society of Photo-Optical Instrumentation Engineers (SPIE) Conference Series*, Vol. 9151, p. 1, doi:10.1117/12.2057108.

Ellis, S. C., Min, S. S., Leon-Saval, S. G., *et al.* (2018). On the origin of core-to-core variations in multi-core fibre Bragg gratings, *Proceedings of SPIE* **10706**, p. 107064U.

Ellis, S. C., Tinney, C. G., Burgasser, A. J., *et al.* (2005). The 2MASS wide-field T dwarf search. V. Discovery of a T dwarf via methane imaging, *AJ* **130**, 5, p. 2347.

Ellis, S. C., Kuhlmann, S., Kuehn, K., *et al.* (2017). Photonic ring resonator filters for astronomical OH suppression, *Optics Express* **25**, p. 15868, doi:10.1364/OE.25.015868.

Farsari, M., Piqué, A., and Sugioka, K. (2019). Laser writing: Feature introduction, *Optical Materials Express* **9**, 11, pp. 4237–4238, doi:10.1364/OME.9.004237, http://www.osapublishing.org/ome/abstract.cfm?URI=ome-9-11-4237.

Ferrari, S., Schuck, C., and Pernice, W. (2018). Waveguide-integrated superconducting nanowire single-photon detectors, *Nanophotonics* **7**, p. 1725.

Fleischer, H. (2011). Vibro-acoustic measurements on the violoncello, in *Analysis and Description of Music Instruments using Engineering Methods*, p. 115.

Fontaine, N. K., Ryf, R., Chen, H., *et al.* (2019). Laguerre-Gaussian mode sorter, *Nature Communications* **10**, 1865, doi:10.1038/s41467-019-09840-4, arXiv:1803.04126 [physics.optics].

Fried, D. L. (1978). Probability of getting a lucky short-exposure image through turbulence, *Journal of the Optical Society of America (1917–1983)* **68**, pp. 1651–1658.

Froehly, C. (1981). Coherence and interferometry through optical fibers, in M. H. Ulrich and K. Kjaer (eds.), *Scientific Importance of High Angular Resolution at Infrared and Optical Wavelengths*, pp. 285–293.

Fuerbach, A., Bharathan, G., and Ams, M. (2019). Grating Inscription Into Fluoride Fibers: A Review, *IEEE Photonics Journal* **11**, 5, p. 2940249, doi:10.1109/JPHOT.2019.2940249.

Gaia Collaboration *et al.* (2016). The Gaia mission, *A&A* **595**, A1, doi:10.1051/0004-6361/201629272, arXiv:1609.04153 [astro-ph.IM].

Gatkine, P., Veilleux, S., Hu, Y., *et al.* (2016). Development of high-resolution arrayed waveguide grating spectrometers for astronomical applications: First results, *Proceedings of SPIE* **9912**, 991271, doi:10.1117/12.2231873.

Gatkine, P., Veilleux, S., Hu, Y., *et al.* (2017). Arrayed waveguide grating spectrometers for astronomical applications: New results, *Optics Express* **25**, p. 17918, doi:10.1364/OE.25.017918.

Gillingham, P. R. and Jones, D. J. (2000). Optical design for IRIS2: the AAT's next infrared spectrometer, *Proc. SPIE* **4008**, p. 1084.

Gillingham, P. R., Miziarski, S., Akiyama, M., *et al.* (2000). Echidna: A multifiber positioner for the Subaru prime focus, *Proceedings of SPIE*, pp. 1395–1403.

Glauber, R. J. (2006). Nobel lecture: One hundred years of light quanta, *Reviews of Modern Physics* **78**, p. 1267.

Goebel, T. A., Bharathan, G., Ams, M., *et al.* (2018). Realization of aperiodic fiber Bragg gratings with ultrashort laser pulses and the line-by-line technique, *Optics Letters* **43**, 15, p. 3794, doi:10.1364/OL.43.003794.

Goodwin, M., Heijmans, J., Saunders, I., *et al.* (2010). Starbugs: Focal plane fiber positioning technology, *Proceedings of SPIE* **7739**, 77391E, doi:10.1117/12.856777.

Gravity Collaboration *et al.* (2017). First light for GRAVITY: Phase referencing optical interferometry for the very large telescope interferometer, *A&A* **602**, A94, doi:10.1051/0004-6361/201730838, arXiv:1705.02345 [astro-ph.IM].

Gravity Collaboration *et al.* (2018). Detection of orbital motions near the last stable circular orbit of the massive black hole SgrA*, *A&A* **618**, L10, doi:10.1051/0004-6361/201834294, arXiv:1810.12641.

Gravity Collaboration *et al.* (2019). Test of Einstein equivalence principle near the Galactic center supermassive black hole, *arXiv e-prints* arXiv:1902.04193.

Gris–Sánchez, I., Van Ras, D., and Birks, T. A. (2016). The Airy fiber: An optical fiber that guides light diffracted by a circular aperture, *Optica* **3**, 3, p. 270, doi:10.1364/OPTICA.3.000270.

Guerin, W., Rivet, J. P., Fouché, M., *et al.* (2018). Spatial intensity interferometry on three bright stars, *MNRAS* **480**, 1, pp. 245–250, doi:10.1093/mnras/sty1792, arXiv:1805.06653 [astro-ph.IM].

Guyon, O., Pluzhnik, E. A., Galicher, R., *et al.* (2005). Exoplanet imaging with a phase-induced amplitude apodization coronagraph. I. Principle, *Astrophysics Journal* **622**, 1, pp. 744–758, doi:10.1086/427771, arXiv:astro-ph/0412179 [astro-ph].

Halir, R., Bock, P. J., Cheben, P., *et al.* (2015). Waveguide sub-wavelength structures: A review of principles and applications, *Laser & Photonics Review* **9**, 1, pp. 25–49, doi:10.1002/lpor.201400083.

Halverson, S., Mahadevan, S., Ramsey, L., *et al.* (2014). Development of fiber Fabry-Perot interferometers as stable near-infrared calibration sources for high resolution spectrographs, *PASP* **126**, 939, p. 445, doi:10.1086/676649, arXiv:1403.6841 [astro-ph.IM].

Hanbury Brown, R. and Twiss, R. Q. (1956). Correlation between photons in two coherent beams of light, *Nature* **177**, p. 27.

Hardy, J. W. (1998). *Adaptive Optics for Astronomical Telescopes* (Oxford University Press).

Harwit, M. (2003). Photon orbital angular momentum in astrophysics, *Astrophysics Journal* **597**, 2, pp. 1266–1270, doi:10.1086/378623, arXiv:astro-ph/0307430 [astro-ph].

Haynes, R., McNamara, P., Marcel, J., *et al.* (2006). Advances in infrared and imaging fibres for astronomical instrumentation, in *Society of Photo-Optical Instrumentation Engineers (SPIE) Conference Series, Proceedings of SPIE*, Vol. 6273, p. 62733U, doi:10.1117/12.671025, astro-ph/0606295.

Henein, S., Spanoudakis, P., Schwab, P., *et al.* (2004). Mechanical slit mask mechanism for the James Webb Space Telescope spectrometer, *Proceedings of SPIE* **5487**, pp. 765–776, doi:10.1117/12.551106.

Hill, J. M., Angel, J. R. P., Scott, J. S., *et al.* (1980). Multiple object spectroscopy - The Medusa spectrograph, *Astrophysics Journal* **242**, pp. L69–L72, doi:10.1086/183405.

Höpker, J. P., Gerrits, T., Lita, A., *et al.* (2019). Integrated transition edge sensors on titanium in-diffused lithium niobate waveguides, *APL Photonics* **4**, p. 056103.

Horton, A. and Bland–Hawthorn, J. (2007). Coupling light into few-mode optical fibres I: The diffraction limit, *Optics Express* **15**, p. 1443.

Horton, A., Content, R., Ellis, S., *et al.* (2014). Photonic lantern behaviour and implications for instrument design, in *Society of Photo-Optical Instrumentation Engineers (SPIE) Conference Series, Society of Photo-Optical Instrumentation Engineers (SPIE) Conference Series*, Vol. 9151, p. 22, doi:10.1117/12.2054570, arXiv:1407.4191 [astro-ph.IM].

Huang, G., Park, T.-H., and Oh, M.-C. (2017). Broadband integrated optic polarization splitters by incorporating polarization mode extracting waveguide, *Scientific Reports* **7**, 4789, doi:10.1038/s41598-017-05324-x.

Hutley, M. C. (1982). *Diffraction Gratings* (Elsevier Science).

Iwamuro, F., Maihara, T., Oya, S., *et al.* (1994). Development of an OH-airglow suppressor spectrograph, *PASJ* **46**, p. 515.

Iwamuro, F., Motohara, K., Maihara, T., *et al.* (2001). OHS: OH-Airglow Suppressor for the Subaru Telescope, *PASJ* **53**, p. 355, arXiv:astro-ph/0101076.

Javan, A., Bennett, W. R., and Herriot, D. R. (1961). Population inversion and continuous optical maser oscillation in a gas discharge containing a He-Ne mixture, *Physical Review Letter* **6**, p. 106.

Jocou, L., Perraut, K., Moulin, T., *et al.* (2014). The beam combiners of Gravity VLTI instrument: Concept, development, and performance in laboratory, *Proceedings of SPIE* **9146**, p. 91461J.

Johnston, K. J. and de Vegt, C. (1999). Reference frames in astronomy, *ARA&A* **37**, 1, pp. 97–125.

Jovanovic, N., Schwab, C., Guyon, O., *et al.* (2017). Efficient injection from large telescopes into single-mode fibres: Enabling the era of ultra-precision astronomy, *A&A* **604**, A122, doi:10.1051/0004-6361/201630351, arXiv:1706.08821 [astro-ph.IM].

Jovanovic, N., Spaleniak, I., Gross, S., *et al.* (2012a). Integrated photonic building blocks for next-generation astronomical instrumentation I: The multimode waveguide, *Optics Express* **20**, p. 17029, doi:10.1364/OE.20.017029.

Jovanovic, N., Tuthill, P. G., Norris, B., *et al.* (2012b). Starlight demonstration of the Dragonfly instrument: An integrated photonic pupil-remapping interferometer for high-contrast imaging, *MNRAS* **427**, 1, pp. 806–815, doi:10.1111/j.1365-2966.2012.21997.x, arXiv:1210.0603 [astro-ph.IM].

Kern, P., Malbet, F., Schanen-Duport, I., *et al.* (1997). Integrated optics single-mode interferometric beam combiner for near infrared astronomy, in P. Kern and F. Malbet (eds.), *Integrated Optics for Astronomical Interferometry*, p. 195.

Kitchin, C. R. (2021). *Astrophysical Techniques*, 7th Edition (CRC Press).

Kleppner, D. (2008). Hanbury Brown's steamroller, *Physics Today* **61**, 8, p. 8, doi:10.1063/1.2970223.

Kuhlmann, S., Liu, P., Ellis, S. C., *et al.* (2018). Photonic ring resonator notch filters for astronomical OH suppression, in *Nanophotonics Australasia 2017, Society of Photo-Optical Instrumentation Engineers (SPIE) Conference Series*, Vol. 10456, p. 104564B, doi:10.1117/12.2283353.

Kumar, A., Thyagarajan, K., and Ghatak, A. K. (1983). Analysis of rectangular-core dielectric waveguides? An accurate perturbation approach, *Optics Letter* **8**, p. 63.

Labeyrie, A., Lipson, S. G., and Nisenson, P. (2014). *An Introduction to Optical Stellar Interferometry* (Cambridge University Press).

Le Bouquin, J. B., Berger, J. P., Lazareff, B., *et al.* (2011). PIONIER: A 4-telescope visitor instrument at VLTI, *A&A* **535**, A67, doi:10.1051/0004-6361/201117586, arXiv:1109.1918 [astro-ph.IM].

Le Coarer, E., Blaize, S., Benech, P., *et al.* (2007). Wavelength-scale stationary-wave integrated Fourier-transform spectrometry, *Nature Photonics* **1**, pp. 473–478, doi:10.1038/nphoton.2007.138, arXiv:0708.0272 [physics.optics].

Lee, C., Chu, S. T., Little, B. E., *et al.* (2012). Portable frequency combs for optical frequency metrology, *Optics Express* **20**, p. 16671, doi:10.1364/OE.20.016671.

Leon-Saval, S. G., Birks, T., Bland-Hawthorn, J., *et al.* (2005). Multimode fiber devices with single-mode performance, *Optics Letters* **30**, p. 19.

Leon-Saval, S. G., Betters, C. H., Salazar-Gil, J. R., *et al.* (2017). Divide and conquer: An efficient solution to highly multimoded photonic lanterns from multicore fibres, *Optics Express* **25**, p. 17530, doi:10.1364/OE.25.017530.

Lewis, I. J., Cannon, R. D., Taylor, K., *et al.* (2002). The Anglo-Australian observatory 2dF facility, *mn* **333**, pp. 279–299, doi:10.1046/j.1365-8711.2002.05333.x, astro-ph/0202175.

Li, C., Liu, D., and Dai, D. (2019). Multimode silicon photonics, *Nanophotonics* **8**, p. 227.

Lindley, E. (2017). *Multicore Fibre Bragg Gratings for Astronomy*, Ph.D. thesis, The University of Sydney.

Lindley, E., Min, S.-S., Leon-Saval, S., *et al.* (2014). Demonstration of uniform multicore fiber Bragg gratings, *Optics Express* **22**, p. 31575, doi:10.1364/OE.22.031575.

Lippmann, G. (1891). La photographie des couleurs, *CRAS (Paris)* **112**, p. 274.

Lippmann, G. (1894). Sur la théorie de la photographie des couleurs simples et composées, par la méthode interférentielle, *CRAS (Paris)* **118**, p. 92.

Liu, P., Czaplewski, D., Ellis, S., *et al.* (2021). Optimizing photonic ring-resonator filters for OH-suppressed near-infrared astronomy, *Applied Optics* **60**, 13, p. 3865.

Livingston, W. (1993). Selected papers on instrumentation in astronomy, *SPIE Milestone Series* **87**.

Luo, L.-W., Ophir, N., Chen, C. P., *et al.* (2014). Wdm-compatible mode-division multiplexing on a silicon chip, *Nature Communications* **5**, p. 3069.

Maihara, T. and Iwamuro, F. (2000). An OH airglow supression spectrograph with multi-object feeder, in W. van Breugel and J. Bland-Hawthorn (eds.), *Imaging the Universe in Three Dimensions, Astronomical Society of the Pacific Conference Series*, Vol. 195, p. 585.

Maihara, T., Iwamuro, F., Hall, D. N., *et al.* (1993). OH airglow suppressor spectrograph: design and prospects, in A. M. Fowler (ed.), *Proc. SPIE Vol. 1946, p. 581-586, Infrared Detectors and Instrumentation, Presented at the Society of Photo-Optical Instrumentation Engineers (SPIE) Conference*, Vol. 1946, p. 581.

Marcatili, E. A. (1969). Dielectric rectangular waveguide and directional coupler for integrated optics, *Bell System Technical Journal* **48**, p. 2071.

Marr, J. M., Snell, R. L., and Kurtz, S. E. (2015). *Fundamentals of Radio Astronomy: Observational Methods* (CRC Press).

Marshall, G. D., Ams, M., and Withford, M. J. (2006). Direct laser written waveguide-Bragg gratings in bulk fused silica, *Optics Letters* **31**, pp. 2690–2691, doi:10.1364/OL.31.002690.

Maughan, B. J., Ellis, S. C., Jones, L. R., *et al.* (2006). XMM-Newton observes Cl J0152.7-1357: A massive galaxy cluster forming at merger crossroads at z = 0.83, *Astrophysics Journal* **640**, p. 219.

Mawet, D., Serabyn, E., Liewer, K., *et al.* (2009). Optical vectorial vortex coronagraphs using liquid crystal polymers: Theory, manufacturing and laboratory demonstration, *Optics Express* **17**, 3, pp. 1902–1918, doi:10.1364/OE.17.001902, arXiv:0912.0311 [astro-ph.IM].

Mawet, D., Serabyn, E., Liewer, K., *et al.* (2010). The Vector Vortex Coronagraph: Laboratory Results and First Light at Palomar Observatory, *Astrophys. J.* **709**, 1, pp. 53–57, doi:10.1088/0004-637X/709/1/53, arXiv:0912.2287 [astro-ph.IM].

McLean, I. S. (2008). *Electronic Imaging in Astronomy: Detectors and Instrumentation*, (2nd edition) (Praxis Publishing).

McLean, I. S., Steidel, C. C., Epps, H., *et al.* (2010). Design and development of MOSFIRE: The multi-object spectrometer for infrared exploration at the Keck Observatory, *Proceedings of SPIE* **7735**, 77351E, doi:10.1117/12.856715.

Midwinter, J. E. (1975). The prism-taper coupler for the excitation of single modes in optical transmission fibres, *Optics Quantum Electronics* **7**, p. 297.

Miller, D. A. B. (2013). Self-aligning universal beam coupler, *Optics Express* **21**, 5, pp. 6360–6370.

Minardi, S., Harris, R. J., and Labadie, L. (2021). Astrophotonics: Astronomy and modern optics, *The Astronomy and Astrophysics Review* **29**, 1, 6, doi:10.1007/s00159-021-00134-7, arXiv:2003.12485 [astro-ph.IM].

Mosley, P. J., Gris-Sánchez, I., Stone, J. M., *et al.* (2014). Characterizing the variation of propagation constants in multicore fiber, *Optics Express* **22**, p. 25689, doi:10.1364/OE.22.025689.

Motohara, K., Iwamuro, F., Maihara, T., *et al.* (2002). CISCO: Cooled infrared spectrograph and camera for OHS on the subaru telescope, *PASJ* **54**, p. 315.

Murphy, M. T., Udem, T., Holzwarth, R., *et al.* (2007). High-precision wavelength calibration of astronomical spectrographs with laser frequency combs, *MNRAS* **380**, 2, pp. 839–847, doi:10.1111/j.1365-2966.2007.12147.x, arXiv:astro-ph/0703622 [astro-ph].

Neo, R., Goodwin, M., Zheng, J., *et al.* (2016). Measurement and limitations of optical orbital angular momentum through corrected atmospheric turbulence, *Optics Express* **24**, 3, p. 2919, doi:10.1364/OE.24.002919.

Nguyen, H. T., Zemcov, M., Battle, J., *et al.* (2016). Spatial and temporal stability of airglow measured in the Meinel band window at 1191.3 nm, *PASP* **128**, 9, p. 094504, doi:10.1088/1538-3873/128/967/094504, arXiv:1510.07567 [astro-ph.IM].

Norris, B. and Bland-Hawthorn, J. (2019). Astrophotonics: The rise of integrated photonics in astronomy, *Optics & Photonics News* **30**, 5, p. 26, doi:10.1364/OPN.30.5.000026, arXiv:1909.10688 [astro-ph.IM].

Norris, B., Cvetojevic, N., Gross, S., *et al.* (2014). High-performance 3D waveguide architecture for astronomical pupil-remapping interferometry, *Optics Express* **22**, p. 18335, doi:10.1364/OE.22.018335, arXiv:1405.7428 [astro-ph.IM].

Norris, B., Schworer, G., Tuthill, P., *et al.* (2015). The VAMPIRES instrument: Imaging the innermost regions of protoplanetary discs with polarimetric interferometry, *MNRAS* **447**, pp. 2894–2906, doi:10.1093/mnras/stu2529.

Norris, B. R. M. (2018). Astrophotonic interferometry: Coherently moulding the flow of starlight, *Proceedings of SPIE* **10701**, p. 107011Q.

Norris, B. R. M., Wei, J., Betters, C. H., *et al.* (2020). An all-photonic focal-plane wavefront sensor, *Nature Communications* **11**, 5335, doi:10.1038/s41467-020-19117-w, arXiv:2003.05158 [astro-ph.IM].

O'Brien, K. (2020). Kidspec: An mkid-based medium-resolution, integral field spectrograph, *Journal of Low Temperature Physics* doi:10.1007/s10909-020-02347-z, https://doi.org/10.1007/s10909-020-02347-z.

Obrzud, E., Rainer, M., Harutyunyan, A., *et al.* (2019). A microphotonic astrocomb, *Nature Photonics* **13**, pp. 31–35.

Offer, A. R. and Bland–Hawthorn, J. (1998). Rugate filters for OH-suppressed imaging at near-infrared wavelengths, *MNRAS* **299**, p. 176, astro-ph/9707298.

Okamoto, K. (2006). *Fundamentals of Optical Waveguides* (Elsevier).

Oliva, E., Origlia, L., Scuderi, S., *et al.* (2015). Lines and continuum sky emission in the near infrared: Observational constraints from deep high spectral resolution spectra with GIANO-TNG, *A&A* **581**, A47, doi:10.1051/0004-6361/201526291, arXiv:1506.09004 [astro-ph.IM].

Pathak, S., Dumon, P., Van Thourhout, D., *et al.* (2014). Comparison of AWGs and Echelle gratings for wavelength division multiplexing on silicon-on-insulator, *IEEE Photonics Journal* **6**, 2361658, doi:10.1109/JPHOT.2014.2361658.

Purcell, E. M. (1956). The question of correlation between photons in coherent light rays, *Nature* **178**, p. 1449.

Quimby, R. S. (2006). *Photonics and Lasers: An Introduction* (John Wiley & Sons, Inc., Hoboken, New Jersey).

Ramachandran, S. and Parmigiani, F. (2019). Special topic on intermodal and multimode fiber photonics, *APL Photonics* **4**, 7, p. 070401, doi:10.1063/1.5112781.

Robertson, J. G. and Bland-Hawthorn, J. (2012). Compact high-resolution spectrographs for large and extremely large telescopes: Using the diffraction limit, *Proceedings of SPIE* **8446**, 844623, doi: 10.1117/12.924937, arXiv:1208.4667 [astro-ph.IM].

Rousselot, P., Lidman, C., Cuby, J.-G., *et al.* (2000). Night-sky spectral atlas of OH emission lines in the near-infrared, *A&A* **354**, p. 1134.

Russell, P. (2003). Photonic crystal fibers, *Science* **299**, p. 5605.

Saleh, B. E. A. and Teich, M. C. (2007). *Fundamentals of Photonics*, 2nd Edition (John Wiley).

Schroeder, D. V. (2000). *An Introduction to Thermal Physics* (Addison Wesley, San Francisco).

Schwab, C., Stürmer, J., Gurevich, Y. V., *et al.* (2015). Stabilizing a Fabry-Perot Etalon Peak to 3 cm s^{-1} for spectrograph calibration, *PASP* **127**, 955, p. 880, doi:10.1086/682879, arXiv:1404.0004 [astro-ph.IM].

Senior, J. M. (1992). *Optical Fiber Communications: Principles and Practice*, 2nd edition (Prentice-Hall).

Shaklan, S. and Roddier, F. (1988). Coupling starlight into single-mode fiber optics, *Applied Optics* **27**, pp. 2334–2338, doi:10.1364/AO.27.002334.

Sheinis, A. I., Jimenez, B. A., Asplund, M., *et al.* (2015). First light results from the high efficiency and resolution multi-element spectrograph at the Anglo-Australian telescope, *Journal of Astronomical Telescopes, Instruments, and Systems* **1**, 3, pp. 1–18, doi:10.1117/1.JATIS.1.3.035002, https://doi.org/10.1117/1.JATIS.1.3.035002.

Snyder, A. W. and Love, J. D. (1983). *Optical Waveguide Theory* (Chapman and Hall, London and New York).

Somekh, S., Garmire, E., Yariv, A., *et al.* (1973). Channel optical waveguides and directional couplers in GaAs–Imbedded and ridged, *IEEE Journal of Quantum Electronics* **9**, pp. 686–686, doi:10.1109/JQE.1973.1077542.

Spaleniak, I. (2014). *Overcoming the effects of the Earth's atmosphere on astronomical observations with 3D integrated photonic technologies*, Ph.D. thesis, Macquarie University.

Spaleniak, I., Gross, S., Jovanovic, N., *et al.* (2014). Multiband processing of multimode light: combining 3D photonic lanterns with waveguide Bragg gratings, *Laser & Photonics Review* **8**, 1, p. L1.

Spaleniak, I., Jovanovic, N., Gross, S., *et al.* (2012). Enabling photonic technologies for seeing-limited telescopes: Fabrication of integrated photonic lanterns on a chip, in *Society of Photo-Optical Instrumentation Engineers (SPIE) Conference Series, Society of Photo-Optical Instrumentation Engineers (SPIE) Conference Series*, Vol. 8450, p. 15, doi: 10.1117/12.925264.

Spaleniak, I., Jovanovic, N., Gross, S., *et al.* (2013). Integrated photonic building blocks for next-generation astronomical instrumentation II: The multimode to single mode transition, *Optics Express* **21**, p. 27197, doi:10.1364/OE.21.027197, arXiv:1311.0578 [astro-ph.IM].

Steinmetz, T., Wilken, T., Araujo-Hauck, C., *et al.* (2008). Laser frequency combs for astronomical observations, *Science* **321**, p. 1335, arXiv:0809.1663.

Stoll, A., Wang, Y., Madhav, K., *et al.* (2020). Integrated echelle gratings for astrophotonics, in S. C. Ellis and C. d'Orgeville (eds.), *Advances in Optical Astronomical Instrumentation 2019*, Vol. 11203, International Society for Optics and Photonics (SPIE), pp. 61–62, doi:10.1117/12.2541554, https://doi.org/10.1117/12.2541554.

Suh, M.-G., Yi, X., Lai, Y.-H., *et al.* (2019). Searching for exoplanets using a microresonator astrocomb, *Nature Photonics* **13**, pp. 25–30.

Sullivan, P. W. and Simcoe, R. A. (2012). A calibrated measurement of the near-IR continuum sky brightness using magellan/FIRE, *PASP* **124**, pp. 1336–1346, doi:10.1086/668849, arXiv:1207.0817 [astro-ph.IM].

Thomson, R. R., Kar, A. K., and Allington-Smith, J. (2009). Ultrafast laser inscription: An enabling technology for astrophotonics, *Optics Express* **17**, pp. 1963–1969, doi:10.1364/OE.17.001963, arXiv:0908.1325 [astro-ph.IM].

Thomson, R. R., Birks, T., Leon-Saval, S., *et al.* (2011). Ultrafast laser inscription of an integrated photonic lantern, *Optics Express* **19**, pp. 5698–5705, doi:10.1364/OE.19.005698.

Thomson, R. R., Harris, R. J., Birks, T. A., *et al.* (2012). Ultrafast laser inscription of a 121-waveguide fan-out for astrophotonics, *Optics Letters* **37**, p. 2331, doi:10.1364/OL.37.002331, arXiv:1203.4584 [physics.optics].

Tinney, C. G., Ryder, S. D., Ellis, S. C., *et al.* (2004). IRIS2: A working infrared multi-object spectrograph and camera, in A. F. M. Moorwood and M. Iye (eds.), *Ground-based Instrumentation for Astronomy, Society of Photo-Optical Instrumentation Engineers (SPIE) Conference Series*, Vol. 5492, pp. 998–1009, doi:10.1117/12.550980.

Trinh, C. Q., Ellis, S. C., Bland-Hawthorn, J., *et al.* (2013a). The nature of the near-infrared interline sky background using fibre Bragg grating OH suppression, *MNRAS* **432**, pp. 3262–3277, doi:10.1093/mnras/stt677, arXiv:1301.0326 [astro-ph.IM].

Trinh, C. Q., Ellis, S. C., Bland-Hawthorn, J., *et al.* (2013b). GNOSIS: The first instrument to use fiber Bragg gratings for OH suppression, *AJ* **145**, 51, doi:10.1088/0004-6256/145/2/51, arXiv:1212.1201 [astro-ph.IM].

Tuthill, P. (2018). Masking interferometry at 150: Old enough to mellow on redundancy? *Proceedings of SPIE* **10701**, p. 107010S.

Watson, F. G. (1995). Multifiber waveguide spectrograph for astronomy, *Proceedings of SPIE* **2476**, pp. 68–74.

Watson, F. G. (1996). The waveguide spectrograph — A new tool for astrophysics. *Australian Journal of Astronomy* **6**, p. 263.

Westerveld, W. J., Leinders, S. M., van Dongen, K. W. A., *et al.* (2012). Extension of Marcatili's analytical approach for rectangular silicon optical waveguides, *Journal of Lightwave Technology* **30**, pp. 2388–2401, doi:10.1109/JLT.2012.2199464, arXiv:1504.02963 [physics.optics].

Woods, T. N., Wrigley III, R. T., Rottman, G. J., *et al.* (1994). Scattered-light properties of diffraction gratings, *Applied Optics* **33**, p. 4273.

Xie, S., Zhan, J., Hu, Y., *et al.* (2018). Add-drop filter with complex waveguide Bragg grating and multimode interferometer operating on arbitrarily spaced channels, *Optics Letters* **43**, 24, p. 6045, doi: 10.1364/OL.43.006045.

Xing, X., Zhai, C., Du, H., *et al.* (1998). Parallel controllable optical fiber positioning system for LAMOST, *Proceedings of SPIE* **3352**, pp. 839–849, doi:10.1117/12.319309.

Yoffe, G. W., Krug, P. A., Ouellette, F., *et al.* (1995). Passive temperature-compensating package for optical fiber gratings, *Applied Optics* **34**, 30, p. 6859, doi:10.1364/AO.34.006859.

Yu, Y., Ma, F., Luo, X.-Y., *et al.* (2020). Entanglement of two quantum memories via fibres over dozens of kilometres, *Nature* **578**, 7794, p. 240, doi:10.1038/s41586-020-1976-7, arXiv:1903.11284 [quant-ph].

Zhang, H., Eaton, S. M., and Herman, P. R. (2007). Single-step writing of Bragg grating waveguides in fused silica with an externally modulated femtosecond fiber laser, *Optics Letters* **32**, p. 2559, doi: 10.1364/OL.32.002559.

Zhelem, R., Brzeski, J., Case, S., *et al.* (2014). KOALA, a wide-field 1000 element integral-field unit for the Anglo-Australian Telescope: Assembly and commissioning, *Proceedings of SPIE* **9147**, p. 91473K.

Zhong, M., Hedges, M. P., Ahlefeldt, R. L., *et al.* (2015). Optically addressable nuclear spins in a solid with a six-hour coherence time, *Nature* **517**, 7533, pp. 177–180, doi:10.1038/nature14025.

Zhu, T., Hu, Y., Gatkine, P., *et al.* (2016). Arbitrary on-chip optical filter using complex waveguide Bragg gratings, *Applied Physics Letters* **108**, 101104, doi:10.1063/1.4943551.

Zins, G., Lazareff, B., Berger, J. P., *et al.* (2011). PIONIER: A Four-telescope Instrument for the VLTI, *The Messenger* **146**, pp. 12–17.

INDEX

T

W

Z

CPSIA information can be obtained
at www.ICGtesting.com
Printed in the USA
JSHW010553070423
39454JS00002B/26

9 781800 613256